BATS
OF THE WORLD

a Golden Guide® from St. Martin's Press

by
GARY L. GRAHAM, Ph.D.

Illustrated by
FIONA A. REID

D0801956

St. Marti   ...w York

FOREWORD

This is an especially favorable time to write a book about bats because more and more people are becoming interested in these mammals—and for good reason. Bats are exceedingly fascinating, very diverse, and extremely important. Unfortunately, some people unnecessarily fear bats. Others are not aware of the critical role that bats perform in nature and of how easily some bat populations can be destroyed. An increased interest in bats drives a desire for more information, which leads to a better understanding of bats and an enhanced appreciation of their importance. With more appreciation comes a greater commitment to, and involvement in, bat conservation.

I appreciate the help and enthusiasm given by Caroline Greenberg during the initiation of this project and the persistence and patience of Maury Solomon at its completion. Special thanks to Fiona Reid for the exceptional bat art and attention to detail. I would like to thank Paul Robertson, Jackie Bellwood, and especially Tom Kunz for reviewing early drafts of the manuscript. I also thank Merlin Tuttle and Bat Conservation International for use of their photographs and assistance. I dedicate this book to my late mother, Bobbie, and to my two sons, Fory and Jove, who represent the next generation of bat conservationists.

G.L.G.

ISBN 1-58238-134-8

CONTENTS

INTRODUCING BATS

BATS ARE IMPORTANT to our world. Over the last few decades, much has been learned about how bats help keep our environment healthy. Many species of bats, such as those known as flying foxes, are also surprisingly appealing and intelligent. Ironically, though, bats continue to be among the most misunderstood and feared of all our wildlife. This fear and ignorance have contributed to the almost total destruction of several bat species and threatened the existence of many others. Such losses can seriously

W.E. Rainey

harm ecosystems and reduce the quality of life for many living things, including humans. But with our help, bats can continue to function beneficially in nature.

BATS ARE MAMMALS, and like all other mammals, the females possess mammary glands, where milk is produced and fed to the young. Baby bats, called pups, are born alive and have to be taken care of for an extended period of time. The body of a bat (but not the wings) is covered by hair, as it is in most other mammals.

BATS FLY, which makes them unique among mammals. Although flying squirrels and flying lemurs have names that seem to indicate otherwise, these animals cannot fly. They do not have wings, and thus they can only glide. Many of the distinctive characteristics of bats are a result of their evolutionary specializations for flight.

Bats belong to the order Chiroptera, which means "hand-wing." Species in this order are divided into two suborders: Megachiroptera, which includes the various species of flying foxes, and Microchiroptera. Flying foxes have foxlike faces with large eyes. Most flying foxes do not use echolocation, a kind of natural sonar for locating prey and other objects. Megachiropteran bats are found only in the Old World (Europe, Asia, etc.). Microchiropteran bats do echolocate and are a much more diverse and widely distributed group.

Samoan Flying Fox;
to 13 in. (330 mm);
wingspan of 3 to 4 ft.
(1,000 mm).

FACT AND FICTION

BATS ARE INDEED BENEFICIAL TO US, although this is a notion that some people find difficult to accept. In many parts of the world, even today, bats are hated and feared, and they are often associated with evil and death. This is unfortunate.

Perhaps the reason for these attitudes is that most bats are small and are active mostly at night, when they are difficult to observe. People tend to fear what they do not fully understand. In contrast, where bats are large and can easily be seen during the day, they are seldom feared and are often even highly regarded.

For example, in Chinese folklore the bat has few rivals as a symbol of good luck and good fortune. In fact, the Chinese word for bat, *fu*, is also the Chinese word for good fortune. Five bats are frequently shown together to represent the five blessings—a long life, wealth, good health, love of virtue, and a natural death.

On the Pacific islands of Samoa, flying foxes are depicted as heroes in a folktale where they save a princess. Folktales from many native cultures and ancient societies, including those of Aesop, the famous Greek storyteller, depict conflict or competition between mammals and birds to explain the nocturnal nature of bats.

Many people believe that bats are blind. This is not true. All bats can see, and many have excellent vision.

Bats are often thought of as flying mice. In fact, the German word for bat is *fledermaus*, which means "flying mouse." Most people are surprised when they learn that bats are actually more closely related to primates, including humans, than they are to mice.

People often use the term "dingbat" to imply that someone is not very smart. Bats, however, are actually quite intelligent and under certain special circumstances can be trained.

Bats are often mistakenly depicted as dirty. However, they spend a great deal of time grooming themselves and keeping themselves clean.

Misinformed people think of bats as symbols of evil and associates of Dracula. But in reality, most bats are very gentle animals that will bite only if they are frightened or improperly handled.

As predators of nocturnal insects, pollinators of flowers, and dispersers of tropical plants (by scattering their seeds), bats are crucial to many of the ecosystems upon which we depend. Bats are also very important to the economies of some developing countries. There, many commercially important trees may owe their very existence to bats.

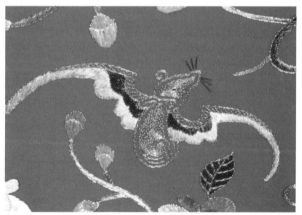

Merlin D. Tuttle, Bat Conservation International

Bats have traditionally been used as a symbol of good luck and good fortune in Chinese art. Here, the bat is depicted on a tapestry mat from the Quing Dynasty.

DIVERSITY

ALMOST 1,000 SPECIES OF BATS are found around the world. This represents nearly a quarter of all known mammals. Rodents are the only mammals that have more species. Such impressive diversity is certainly matched by the tremendous variation in ecology and behavior displayed by bats. Bats eat everything from insects and fruits to nectar, fish, meat (small land vertebrates), and even blood.

Bats are often compared with birds because both of them can fly. However, there are about eight times more species of birds than there are of bats. This may be because bats are a younger group than birds and have had less time to evolve new species. Another reason may be anatomy. In bats the legs, which are part of the wings, cannot be used effectively during various food-gathering activities, such as swimming, diving, running, and digging. In birds the wings are separate from the legs. Thus, many species of birds that use their legs for food gathering have evolved.

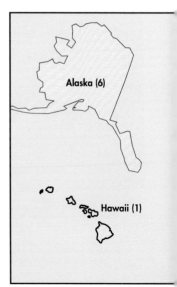

BATS CAN BE FOUND in almost all habitats except extremely hot deserts and the cold polar regions. Species are more diverse in the warmer latitudes. For instance, only one insect-eating species can be found in northern Canada, whereas approximately 150 species are found, along with several non-insect-eating bat species, in some tropical areas near the equator. A similar increase in the number of species and individuals can be observed as one travels down tropical mountainsides from the alpine tundra to lowland rainforest habitats. Bat diversity on islands also decreases as the distance to nearby continents increases. For instance, five species of bats are known to exist in Fiji and three in American Samoa. Only one is native to Hawaii.

In the United States the areas with the greatest diversity of bats are in the Southwest—for instance, in the Chiricahua Mountains of Arizona and Big Bend National Park in Texas. These regions have both diversified habitats and many months when food is abundant.

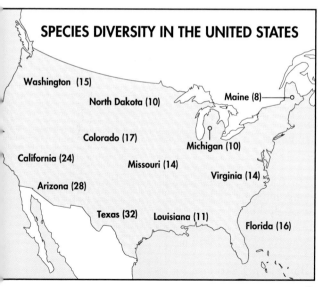

SPECIES DIVERSITY IN THE UNITED STATES

Washington (15)

North Dakota (10)

Maine (8)

Colorado (17)

Michigan (10)

California (24)

Missouri (14)

Virginia (14)

Arizona (28)

Texas (32) Louisiana (11)

Florida (16)

EVOLUTION

THE FOSSIL RECORD for bats is far from complete. This is partly because bats are small and have delicate bones that seldom become fossils. The bones are difficult to recognize even when they are fossilized. Thus, the origin and evolution of these mammals are poorly understood.

The oldest bat fossil, *Icaronycteris index*, became extinct about 60 million years ago in what is now North America. Early bats were insect eaters, as indicated by their teeth and by the fossilized insects found in the stomach of one ancient bat. Megachiropteran bats first appear in the fossil record about 35 million years ago.

The abilities to fly and to catch insects in the dark were important developments in bat evolution. Many scientists

Possible stages in the evolution of flight in bats.

1. Arboreal insectivore

2. Primitive glider

believe that bat ancestors were small, shrewlike mammals that chased flying insects among the leaves of trees and evolved limb membranes that enabled them to glide from branch to branch. The transition from a fixed limb, like that of a flying squirrel, to a movable wing was a critical step in the development of bat flight and evolution. It allowed bats to pursue prey above the trees.

Scientists also suspect the early ancestors of bats were able to echolocate, which enhanced the capture of insects at night. However, the exact origin of echolocation in bats is not known. The type of echolocation used by the microchiropteran bats is not found among flying foxes. This difference, among others, has led some scientists to question the supposedly close relationship between these two particular groups.

4. Bat

3. Movable wings

11

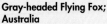
Gray-headed Flying Fox;
Australia

Flying Lemur;
Philippine Islands

The faces of flying foxes resemble those of mammals in three other orders: flying lemurs, tree shrews, and primates.

THE RELATIONSHIP BETWEEN FLYING FOXES AND OTHER BATS has been debated by scientists and other experts for over ten years, yet it remains basically unresolved. Some scientists think that flying foxes and primates are more closely related to each other than flying foxes are to other bats because they share some characteristics not found in microchiropteran bats. These characteristics include the nerve pathways associated with vision, part of the male reproductive system, and an inability to echolocate.

Common Tree Shrew

**Ring-tailed Lemur;
Madagascar**

Such a view implies that the similarities, including flight, shared between the two bat groups evolved two times in mammals. Recently, however, a number of studies using new genetic and analytical technologies dispute this view.

Most, but not all, scientists who study the evolutionary relationships between animals now conclude that megachiropteran and microchiropteran bats are much more closely related to each other than either group is to primates. This means that all bats belong to the same order and that flight evolved only once.

13

FLIGHT

TRUE SUSTAINED FLIGHT distinguishes bats from all other mammals and is a trait shared only with birds and insects. To those creatures that possess it, flight offers numerous advantages. It allows access to additional food resources, as well as to roosting and nesting sites that are unavailable to flightless animals. Flight makes long-distance migrations possible and allows animals to get past barriers that are difficult for nonflying animals. In addition, a flying bat or bird uses less than a quarter of the energy consumed by similarly sized land species moving over the same distance. Finally, flight is also an effective way to escape from most predators.

Flight has thus been one of the principal factors in the evolutionary success of bats and birds. The reason flight has not evolved among other vertebrates (animals with a backbone) may have to do with the many complex specializations required for flight.

THE BASIC REQUIREMENTS for flight include a method for maintaining the body above the ground (lift), a means of propulsion through the air (thrust), and a streamlined body to cut down on air resistance. The bat's wing, which provides both lift and thrust, is truly a masterpiece of evolutionary design. It incorporates the same basic arm and hand bones found in humans and most other mammals, except that in bats the hand and finger bones are very long and slender and there are fewer digits. One of the forearm bones, the ulna, is reduced in size.

Flight membranes are very thin sections of skin stretched between the arms, fingers, body, legs, and feet. Although rather delicate-looking, these membranes are even more resistant than rubber gloves to tearing by sharp objects.

14

The muscles that move the wing are located on the chest, back, and shoulder rather than on the wing. This allows the bat to fly with less expenditure of energy. Bat legs are used more for flight than for moving about on land. Hence, the pelvis and legs are reduced in size, which contributes to a slender body shape.

PARTS OF A BAT

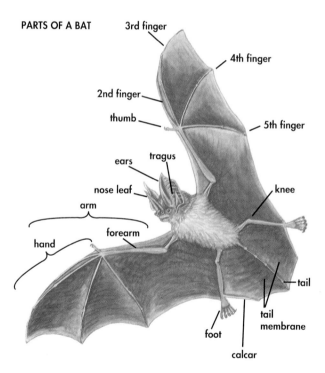

Common Sword-nosed Bat; to 2½ in. (60 mm); found from southern Mexico to Bolivia and southern Brazil.

LIFT AND PROPULSION are achieved by the downstroke of the wing. Lift is caused by air moving faster over the top of the wing than under it. It can be increased by increasing the speed of air moving past the wing, by changing the curvature of the wing, and by adjusting the angle of the downstroke.

The membrane between the body and the fifth finger is the lift-generating part of the wing. Forward thrust is produced by the membrane between the second and the fifth fingers.

Wingbeats are similar to the arm movements of a human swimming the breaststroke. Many species can take off from the ground, but most bats begin flight by taking off from a roosting site. Landing involves decreasing airspeed until the bat "stalls." Most bats grab hold of a branch or other object and then assume an upside-down position. Others perform a flip first and then grab hold.

Relatively short, broad wings, such as those belonging to many species in the family Phyllostomidae, permit highly maneuverable flight in a tight space. Species with long, narrow wings are usually swift flyers but maneuver poorly (for instance, many bats in the family Molossidae).

Most bats have a membrane between their legs. This tail, or interfemoral, membrane increases the overall lifting surface of the bat's wings and may also act as an air brake. It is even used by a number of species to trap insects during flight.

Birds have evolved somewhat different methods to accomplish flight. Perhaps the most obvious difference is the use of feathers rather than membranes. Bird wings are also characterized by the fusion of the bones and their reduced size.

In addition, birds have fewer flight muscles than bats. These muscles are attached to the keel, or ridge, on a bird's sternum. Bats do not have a keeled sternum.

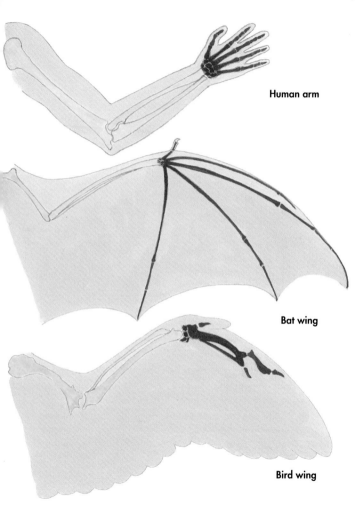

Human arm

Bat wing

Bird wing

The arm bones above are yellow, the hand is red,
and the finger bones are blue.

Top: Black Flying Fox; to 11 in. (280 mm); Sulawesi, New Guinea, Australia. *Bottom:* Salvin's Big-eyed Bat; to 3 in. (75 mm); tropical Mexico south to Peru. Both species have good vision.

ECHOLOCATION AND OTHER SENSES

ALL BATS CAN SEE, but their eyes are better adapted to seeing in the dark. For example, they see objects only in black and white. Microchiropteran bats use their eyes to monitor varying levels of light in the environment and to help them orient themselves in space.

Flying foxes use vision, rather than echolocation, for most of their activities. Consequently, their eyes are large compared with the typically small eyes of echolocating bats. In addition, the eyes of flying foxes possess a reflective layer of tissue under the retina that is typical of most nocturnal mammals. This improves vision by allowing more light to pass over the retina, where there are light-sensitive cells. This reflective layer is not found in other bats, probably as a consequence of their having evolved the ability to echolocate.

The sense of smell is well developed in most bats. It is used to locate and identify certain food items and to recognize roost mates and young. It is also important in the social behavior of some species.

ECHOLOCATION is the process used by some animals to identify, and measure the distance to, objects in the environment. Although this ability, which involves listening to echoes of sounds the echolocator itself produces, is associated mostly with bats, some birds, cetaceans, insectivores, rodents, and even people can echolocate. A person who shouts in a canyon and listens for the echo is using a simple form of echolocation to detect canyon walls.

Humans cannot normally hear bats echolocating. The low-frequency "squeaky" bat sounds that humans hear can also be used by some mother bats to communicate with their young or as warning or recognition calls.

Bats use echolocation to orient themselves in space and to determine the size, shape, texture, distance, speed, and direction of movement of prey or other food items. Some species can detect objects as thin as a human hair. All 800 or so species of microchiropteran bats are thought to use some form of echolocation.

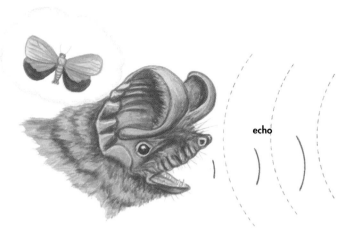

echo

Echolocation pulses emitted by a free-tailed bat bounce off a small moth. The returning echo provides an image of the moth.

SOUND VIBRATIONS form waves in the air, waves that have successive peaks and troughs. Sound frequencies, the number of these wave cycles that occur in a second, are expressed in kilohertz (kHz), one of which equals 1,000 cycles per second. The frequencies used by echolocating bats range from 4 to 210 kHz, but most species use signals between 20 and 80 kHz.

High-frequency, short-wavelength signals are better for identifying small objects than are low-frequency, long-wavelength sounds. Thus, the high-frequency echolocation calls of most bats are more useful for detecting the small insects that make up so much of their diet. Because the maximum frequency usually perceived by humans is around 20 kHz, frequencies above this number are referred to as ultrasonic.

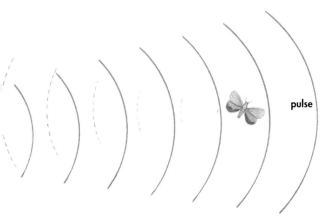

pulse

ECHOLOCATION CALLS, or pulses, are produced in the larynx, or voice box, by forcing air past very thin vocal membranes that are unique to bats. Most species emit signals through the mouth, but some may use a complex nose structure, called a nose leaf, to direct signals from the nostrils. Members of the Old World genus *Rousettus,* the only megachiropteran bats that echolocate, use tongue clicks instead of vocal cords to produce sounds.

Usually, echoes are received by large funnel-shaped ears that face forward. Many species have a vertical flap inside each ear, called a tragus, that may help direct incoming echoes. The internal ear parts are similar in structure and function to those of other mammals, except that the eardrum is somewhat thinner and the cochlea is somewhat larger. The cochlea is specialized for frequencies in the ultrasonic range. The nerve impulses that are generated by these sounds are transported to the brain for processing.

THE BAT'S BRAIN processes information and forms images by comparing outgoing pulses with the echoes. The time between when the bat's call is sent and when the echo returns indicates how far away a particular object is. Echoes also convey information about the size of the target. The direction in which the target is moving is determined by the different arrival times of the sound at each ear and also by the intensity of the echoes as they reach the bat's ears.

Constant-frequency (CF) pulses are echolocation calls produced by the bat at a single frequency or in a narrow band of frequencies. A CF pulse, useful for detecting targets, can also help to determine the relative speed of insects by noting changes in frequency as the object moves. Bats can also use such shifts in frequency to obtain information about fluttering prey.

Greater accuracy about the location and physical properties of targets is provided by frequency-modulated (FM) pulses. Many bats emit pulses that have both CF and FM components. Bats can even broaden their bandwidths through the use of harmonic frequencies, which are multiples of the lowest frequency used in a complex echolocation call.

The effective range of echolocation for most bats is 3 to 9 feet but exceeds 30 feet in some species. Range increases with the size of the target and decreases under more humid conditions. Range also varies with the intensity of the calls. Some bats, such as the "whispering bats" in the family Phyllostomidae, expend little energy on their signals. Several "loud" bats in other families emit calls that are comparable to the loud sounds produced by jackhammers. Fortunately, these intense calls are above the frequencies humans can hear.

ORIENTING THEMSELVES AND OBTAINING FOOD in changing environments are challenges faced by all bats. Solutions are as varied as the species. Individual bats alter their echolocation calls to meet new conditions. When insect-eating bats are searching for food, they typically emit sounds at about 10 pulses per second, just enough to identify potential prey. Once an item is detected, the rate increases to 25 to 50 pulses per second. As the bat approaches its prey and the insect is about to be caught, the rate may rise to as high as 200 pulses per second. Higher pulse rates deliver more precise information about the prey and increase the chances of success.

ECHOLOCATION CALLS ARE IMPORTANT for recognition purposes in the severely crowded conditions associated with some roosts. They are particularly important for mother bats that are attempting to locate their young.

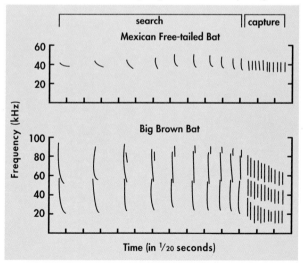

Cave roosts are used by many types of bats.

BAT ROOSTS

ROOSTS ARE IMPORTANT because bats spend over half their lives roosting in a variety of places, from caves to leaves. Roosts provide protection from severe weather, extremes of heat and cold, and predators. They also provide sites for resting, exchanging information, mating, rearing young, hibernating, and digesting food. The number and types of roosts available may influence the diversity of bats found locally.

24

Bats use day roosts principally for extended rest periods. At night, they often use different roosts, for eating and for brief rest periods. Some species hunt for prey from night roosts, and a few tropical bats use roosts as sites for calling to potential mates. In some species, maternity colonies and bachelor groups form and roost apart. They may even inhabit different types of structures. Hibernation roosts, used during cold months, protect bats against freezing temperatures and allow body temperature and metabolism to drop. This saves energy and helps bats survive the winter.

Availability and quality of roosts affect roost choice and loyalty to a particular roost. Individuals may change roosts daily to avoid predators in exposed sites or seasonally to remain close to abundant food supplies. The type of site chosen by individuals is also influenced by the temperature of the surrounding environment, level of moisture in the air, and exposure of the roost to the wind.

Most species use visual and acoustic cues as well as memory to locate roosts. Upon approaching their roost, some species exhibit ritualized behavior. This may involve several passes by the entrance and perhaps a brief landing nearby. During such rituals the bats may exchange information with individuals that have already entered the roost or are in the area.

ROOSTING HABITS vary, often in conjunction with the physical makeup of a species. Bats that roost in crevices, including many free-tailed bats, have flattened skulls and bodies; short, stout legs; short body hair; long hairs on the toes or head; and hips modified so that the legs are extended to the side. These adaptations allow for roosting and crawling in tight places—for instance, under rocks. Bamboo bats crawl through narrow slits into cavities in bamboo plants. Short-tailed bats have special pouches for protecting their wings, plus claws for digging holes in decayed logs to use as roosts. At least 6 genera and 11 species, including those in the families Thyropteridae and Myzopodidae, have thickened pads or suction disks on their thumbs and feet—specializations for attaching themselves to the smooth surfaces of leaves or the slippery walls of bamboo cavities. Several species, such as Red Bats and Hoary Bats, roost in trees in temperate areas; their long woolly fur protects against heat loss.

South American Flat-headed Bat; to 2 in. (55 mm); found from Venezuela to central and eastern Brazil.

Honduran White Bat; to 1¾ in. (45 mm); roosting under a leaf cut to form a tent. This bat is found in the lowlands of Central America.

Most species, however, are not specialized for any particular type of roost. Many species of New World leaf-nosed bats and several Old World fruit bats create tent roosts from the leaves of plants such as palms and bananas. Many species in the tropics roost in foliage or on exposed surfaces such as tree trunks. These species are frequently dark with whitish stripes, which may serve as protective coloration. The pelage (fur) color of others, such as the Eastern Red Bat, may reduce the risk of predation by making the bat look like a dead leaf.

Most bats hang upside down while roosting. This could cause blood to accumulate in the head, but special valves in the veins prevent this. Species that form large roosting colonies—for instance, Mexican Free-tailed Bats—are frequently exposed to levels of ammonia and carbon dioxide in the cave that would be lethal to people. However, these bats have evolved physiological adaptations for surviving under such harsh conditions.

BATS "HANG OUT" in various types of structures, depending on geography and species. Trees are used as roosts more often than anything else, especially in tropical regions, although caves or cavelike structures, including mines and culverts, are the types of roosts most often associated with bats. Other types of roosts include rocky cliffs, animal burrows, termite nests, shrubs with large leaves, and stumps or fallen logs. With trees, different species may roost on trunks, inside large hollow cavities, inside smaller holes, under or among foliage, or under bark. In caves some bat species are found near entrances on walls, while other bats roost on flat ceilings, in ceiling depressions, or inside cracks. Numerous man-made structures used by bats, such as buildings and bridges, offer protection and conditions similar to those of natural roosts. Some species of bats roost alone; other bats roost in groups. Colonies can vary in size from small to very large. A bat may be in close contact with the roost surface or hang from it.

Only roosting "camp" of Marianas Flying Fox left on Guam.

Merlin D. Tuttle, Bat Conservation International

Temple ruins in Tikal, Guatemala, where Hairy-legged Vampire Bats can often be found.

Seminole Bat in Spanish moss.

Termite nests, a popular roosting spot for D'Orbigny's Round-eared Bats.

Lesser White-lined Sac-winged Bats on the trunk of a fig tree. Jamaican Fruit Bats live in the tree's hollow base.

FOOD HABITS

BATS CONSUME an impressive variety of foods—from scorpions, insects, pollen, fruits, and leaves to blood, fish, frogs, birds, and even other bats. The use of such a wide variety of foods is a consequence of the large number of bat species and their foraging specializations. Bats also need to consume very large amounts of food to meet their dietary and energy needs, with most species eating about half their body weight each night. Some fruit-eating bats may devour more than twice their weight during a single night. Undigested food begins to be excreted in most bats about 20 minutes after a meal is eaten.

Because bats are so abundant and diverse, and because they consume such large quantities of food, their foraging activities are crucial to the health of ecosystems around the world. Many species of bats are the principal predators of nocturnal insects or are vital to the pollination and the dispersal of many tropical plants.

CARNIVORY, or the eating of meat, is restricted to several large species, members of the families Nycteridae, Megadermatidae, and Phyllostomidae. These bats capture and eat lizards, frogs, birds, rodents, and other bats. Some species also consume insects and fruits. Carnivorous bats typically bite into the head of their prey, which kills quickly.

Carnivorous bats use both vision and echolocation to find their prey. Some species may locate other bats by actually listening for their echolocation calls. Fringe-lipped Bats often find frogs and katydids by homing in on their mating calls. The more a frog or an insect calls for a mate, the greater the probability it will be detected and eaten by a bat. Some species of frogs and katydids have evolved elaborate mechanisms to protect themselves from these bats.

Fringe-lipped Bat; to 3½ in. (90 mm); with its frog prey. Range is southern Mexico to Peru and southeastern Brazil.

Mexican Fishing Bat; to 3 in. (75 mm); found on islands and along coastal areas of Baja California and adjacent Mexico.

PISCIVORY, or the eating of fish, is a specialized form of meat eating and is used by only about six different species of bats. Bats that catch fish have long legs and large feet with long, sharp, curved claws. Their claws, toes, and legs are flattened, which reduces resistance when they are dragged through water.

These bats forage by flying low over quiet water. They use echolocation to detect the fins of fish or the ripples created by fish swimming near the surface. After finding a fish, the bat drags its talons through the water, grabbing its unsuspecting prey. The captured fish is quickly lifted out of the water and transferred to the bat's mouth. Fish are either eaten on the wing or taken to a perch. Bats have been observed catching 30 to 40 fish per night in a stocked pond.

SANGUIVORY, or feeding on blood, is an unusual dietary habit. It is restricted to just three species, all in the family Phyllostomidae. Common Vampire Bats, the most common of the vampire bats, feed on the blood of mammals. Hairy-legged Vampires consume bird blood, and White-winged Vampire Bats prey on the blood of both birds and mammals. Sanguivorous bats are found only in the New World, from Mexico to the southern tip of South America.

Vampires land on their prey and then try to locate an area of skin to bite that has a large supply of blood near the surface. Heat-sensitive pits on the bat's face help it to determine a choice spot. The bat uses its razor-sharp incisor teeth to make a small, V-shaped cut. The blood is kept flowing by a chemical in the bat's saliva that prevents blood from clotting. Vampires do not suck blood; rather, they lick the blood from the open wound.

Common Vampire Bats drink about 1 fluid ounce of blood per night. Because blood is very thick and the volume consumed each night is large, a bat that has just eaten could have trouble flying. However, the kidneys of vampire bats are quite well adapted for the rapid removal of water. The bat starts to urinate almost as soon as it begins to feed. Additionally, vampire bats are well adapted for taking off with a heavy load. They can jump high into the air, using their extra-strong limbs and long thumbs for a boost.

Cattle and other domestic animals are the most common victims of vampire bats. People are sometimes preyed upon by vampire bats, but human blood is not a preferred food.

Vampires can transmit rabies to cattle and have been blamed for occasional rabies outbreaks in Latin America. Unfortunately, attempts to control vampires by destroying bat roosts have killed hundreds of thousands of beneficial bats as well.

INSECTS and their arthropod relatives, such as spiders and scorpions, provide a rich source of energy, protein, and other nutrients for bats. Insects can be found in great abundance throughout most of the world. About 70 percent of all bats eat insects as the main component of their diet. For some of the other species, insects serve as a protein supplement to a fruit or nectar diet.

Insectivorous bats most often eat beetles, moths, flies, and mosquitoes. They also eat cockroaches, termites, crickets, katydids, cicadas, and night-flying ants. Many insects eaten by bats are considered pests by farmers. A few, such as mosquitoes, can spread disease.

Bats capture their insect prey by foraging on the wing, ambushing flying insects from a perch or picking the insects off vegetation or the ground. Some bats use their mouth to seize flying insects. Others use their wings or tail membrane to first trap prey.

Most species are exceptionally efficient at catching insects. For example, in a laboratory study, Little Brown Bats captured as many as 1,200 fruit flies per hour, or 1 every three seconds!

Nearly a half-million pounds of insects are consumed each summer night by a single colony of 20 million Mexican Free-tailed Bats in Texas. The number of insects eaten each night by all bats around the world is almost too large to imagine.

Such heavy predation has led insects to evolve a number of defenses to help them avoid being eaten by bats. Many species of moths, lacewings, crickets, and mantids, for example, have specialized "ears" and can hear the echolocation calls of foraging bats. If the bat is close, the insect will quickly drop into vegetation or fly downward in a spiraling motion. Several moths can produce sounds that may "jam" the echolocation calls of an approaching bat.

D'Orbigny's Round-eared Bat; to 3½ in.
(90 mm); carrying a large katydid. Found from
Central America to northwestern Peru, southern
Brazil, and Paraguay.

PLANT PARTS used by bats include fruit, nectar, pollen, and leaves. Species in the family Pteropodidae, and most members of the family Phyllostomidae, consume plant parts, especially fruit. These bats disperse or pollinate a large number of plants in tropical forests. For example, about 40 percent of the trees on the Pacific island of Guam depend at least to some degree on bats for seed dispersal or pollination. Bats that consume mostly fruit, which provides energy, must eat large quantities to be sure that they meet their protein needs, or they must supplement their diet with insects.

Some flying foxes (and perhaps others) can apparently extract protein from leaves. By chewing on the leaves for a while, the bat releases the liquid portion and ingests it, spitting out the fibrous residue.

FRUGIVOROUS, or fruit-eating, bats normally prefer fruit that is ripe. They are attracted to plants by the smell of ripening fruit, which is eaten on the spot or taken to a night perch. Bats chew and suck the fruit to separate the juices from the fibrous material and the seeds. The teeth of fruit-eating bats, which are relatively broad and flat, are good for crushing fruit. The tongue presses the fruit against the ridged roof of the mouth in order to extract the juice. Several species have teeth that are grooved down the middle and can be used to siphon off the squeezed juice.

Large seeds are discarded, but smaller ones pass through the bat undamaged, to be deposited at sites distant from the parent plant. Some of these seeds will not grow unless they have passed through a bat's gut. Many plants that are the first to grow in a cleared area have come from bats. Without bats, reforestation of cleared tropical lands would occur much more slowly. Thus, bats are crucial to healthy tropical forest ecosystems.

NECTARIVOROUS bats feed mostly on nectar and pollen. Almost all of them belong to a few species in the families Pteropodidae and Phyllostomidae. Nectar- and pollen-feeding bats may periodically eat insects, and fruit eaters may consume flower products on occasion.

Nectar- and pollen-eating bats behave and even look somewhat like hummingbirds. They are small, delicate bats with long snouts. Their long tongue often has a brushlike tip that actually increases the bat's capacity to lap up nectar. Because their diets do not require much chewing, some nectarivorous species lack lower incisors, and the teeth in the cheek area may only be small pegs.

Specialized hairs on these bats trap pollen. Pollination occurs when a bat transports pollen from one flower to another of the same species. Many tropical plants have flowers that attract bats. These flowers usually open at night, jut out from foliage for easy access, are whitish or greenish in color, may have a strong musky or sour odor, and produce large amounts of nectar and pollen.

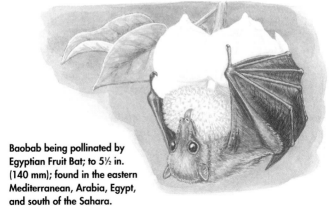

Baobab being pollinated by Egyptian Fruit Bat; to 5½ in. (140 mm); found in the eastern Mediterranean, Arabia, Egypt, and south of the Sahara.

REPRODUCTION

COURTSHIP AND MATING in bats is both complex and interesting. Some species are solitary, whereas others are highly gregarious.

Many species have elaborate mating rituals. Male Greater White-lined Bats maintain roosting territories inside the cavities of large trees for groups of one to eight females. After he is done foraging, the dominant male returns to the roost, where he uses a "chirpy" song to attract females. The male greets each female by hovering over her and singing as she sings. Adjacent males frequently fight over territorial boundaries. They approach each other, wave their wings (each of which has an open scent gland), and occasionally fly at each other. Although male Greater White-lined Bats usually do not injure their adversaries, dominant males in other species actually strike intruders with their wings.

Male Greater White-lined Bat; to 2¼ in. (55 mm); "chirping" to a female. Found from southern Mexico to southeastern Brazil.

Male Hammer-headed Bat; to 11 in. (280 mm);
found in tropical Africa.

The mating ritual of Hammer-headed Bats is called a lek. Nearly 100 males gather at dusk in a narrow band of forest along a river. There, spaced about 30 feet apart, they start singing and flapping their wings rapidly to attract females. The song—a series of deep, froglike honks repeated from 50 to 120 times per minute—is produced by their larynx and amplified by a megaphonelike muzzle. Females move along the line, hovering in front of and inspecting individual males. A female will usually hover several times in front of a prospective mate before landing beside him and copulating briefly.

Male epauleted fruit bats attract mates with a song and dance display during which the males flash large patches of white shoulder fur at the females.

THE REPRODUCTIVE PROCESSES of bats are typical of mammals—with several interesting exceptions. During sexual activity the male mounts his mate from the rear. This typically occurs at a roost or perch. During delivery of a pup, some mothers-to-be reverse their position at the roost so that they are hanging with their heads up. The newborn bat emerges and drops into the membrane between its mother's legs. The mother licks her baby as it climbs up and attaches to a nipple.

The milk of a mother bat is very nutritious, and the pups grow at a rapid rate. Mothers and pups establish a strong bond and can identify each other by odors and by vocal communication—even among millions of roost mates. The young of most smaller species start to fly after three to six weeks and are weaned about a month later. It may take six months or longer for large species, such as flying foxes, to be weaned.

There is wide variation among bats in the timing of reproductive events. The most crucial factor seems to be that young must be weaned when food is abundant. In temperate zones this is the summer season, so bats have only one birth period per year. Tropical bats are exposed to longer periods when food is readily available. Thus most, but not all, give birth twice a year. A few species, such as the Common Vampire Bat, consume food that is available throughout the year, and they can give birth during any season.

Most temperate-zone bats give birth to only one young. Twins occur in a few species, and Eastern Red Bats frequently have quadruplets. Tropical species usually give birth to one baby at a time.

Bats are exceptional in the strategies they have evolved to handle the timing of reproductive events. Hibernating temperate-zone bats usually mate during the fall. The sperm is stored in the female, with fertilization of the egg by the sperm taking place in the spring. Others, such as the Straw-colored Flying Fox, can delay the implantation of a fertilized egg for three to five months. Fertilization and implantation occur soon after copulation in the Jamaican Fruit Bat, but the development of the embryo is delayed for several months. These reproductive strategies are designed to increase the chances that all young are born when food is abundant.

Gestation, or the total time required for an embryo to develop into a baby, varies from less than two months in small plain-nosed bats to as many as eight or nine months in large flying foxes. Gestation for most medium-sized bats requires from 50 to 120 days.

Black-winged Little Yellow Bat; shown here actual size, with newborn attached to nipple *(left)*. **While resting, the baby is hidden beneath the mother's right wing. Ranges from Mexico to northern South America.**

NEWBORN BATS are helpless. They depend on their mothers for food and for teaching them many of the skills needed to survive. Bonding between mother and baby begins at birth as the mother licks and grooms her newborn. She identifies it using smell and vocalizations. Upon returning to the roost, a mother bat can find her own pup, even in a large colony with millions of other bats. Mistakes seldom occur.

A pup being weaned must learn to fly in the dark, sometimes in a crowd of thousands. A mistake could cause the bat to fall to the ground or to the floor of the roost, where danger awaits. Young bats must also learn how to orient themselves and how to obtain food. The mothers assist in this training.

Wahlberg's Epauleted Fruit Bat; to 6 in. (150 mm); watches as its mother snatches a fig and prepares to fly away with it. Its range is Africa south of the Sahara.

BAT BEHAVIOR

MOST BATS ARE INTELLIGENT and can be easily trained, though they should never be kept as pets. They can be taught to fly or walk through mazes and to respond to commands. For instance, Fringe-lipped Bats have been trained by experts in as little as two hours to respond to verbal or hand signals, and to fly toward a frog from a specific direction. Bats in training or rehabilitation learn to recognize and trust their human caretakers. Some species and individuals are more easily trained than others, but most can be trained in a matter of days. In addition to young bats learning from their mothers, adult bats learn from each other. This may help explain how new food resources and new roosting areas can be quickly exploited by a whole colony.

DAILY ACTIVITIES of bats are regulated extensively by light. By being active at night, bats can avoid competition with birds for food resources and can reduce their exposure to predation. Samoan Flying Foxes, however, are active during the day on the Pacific islands of Samoa, where bats have no natural enemies. Many bats change their foraging patterns on bright moonlit nights to reduce the risk of being captured by owls or other predators.

Female Mexican Free-tailed Bats that are nursing young typically hunt insects two times each night for a total of about eight hours in flight. At the other extreme, Heart-nosed Bats spend as little as 11 minutes in foraging flights each night. Instead, they sit on perches and wait for insects or other prey to pass by. Most species are between these two extremes. Although the roost is generally considered a place for resting, activities such as mating, giving birth, and grooming also take place there.

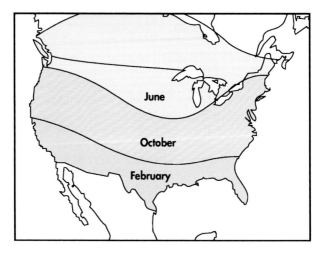

June

October

February

Seasonal migration of Hoary Bats.

MIGRATION is common, due to the combination of cold temperatures and insect scarcity in winter. Most species of bats in northern regions migrate, hibernate, or do both. Northern populations of species that roost in exposed sites usually go south, where some remain active while others hibernate for at least part of the winter. Most bats migrate less than 200 miles, but some species, such as the European Noctule, travel over 2,400 miles (including the return trip). Migratory pathways of bats are less well delineated than those of birds but are similar in that they are in a north-to-south direction. In North America some species migrate along the routes of migratory birds. We know this from evidence of bats and birds that have collided with tall man-made structures located in migratory paths. Although much is still to be discovered, biologists have found out that at least some species use vision and familiar landmarks to orient themselves during migration.

HIBERNATION is another adaptation to winter. Species such as Gray Bats migrate from warm nursery caves, which are suited for the rapid growth of young, to cooler caves with stable winter temperatures. These cooler caves are conducive to hibernation. The body temperature, as well as breathing, metabolism, and blood circulation of a hibernating bat, is reduced to mere survival levels. Bats prepare for hibernation, which may last for many months, by accumulating body fat. They conserve energy by entering torpor (a temporary state of reduced metabolic activity) during the day and foraging at night. Most bats store just enough fat to last through the hibernation period. Any disturbance that causes a bat to awaken and use its fat stores can shorten by a month the time that the animal can hibernate. The hibernation roosts, called hibernacula, usually have very humid conditions; this helps to avoid excessive loss of body water. Bats may spontaneously awaken if the temperature becomes too hot or too cold or if they lose too much water. Hibernacula must also protect against predators.

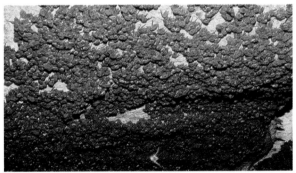

Merlin D. Tuttle, Bat Conservation International

Hibernating Gray Bats.

BAT MORTALITY

MOST BATS DIE while still young. Bats have many natural enemies. Great Horned Owls and several species of hawks snatch Mexican Free-tailed Bats as they depart in columns from their caves in the southwestern United States and Mexico. Bat Hawks and Bat Falcons prey on bats in Africa and Latin America, respectively. Peregrine Falcons eat bats in many parts of the world. On the islands of Fiji, flying foxes make up a large part of the diet of these powerful predators. In tropical areas carnivorous bats eat other bats. Raccoons, house cats, and snakes capture bats as they leave their roosts, and raccoons sometimes snatch hibernating bats from the walls of caves. The average life span of a bat that survives to adulthood is about ten years, but one bat is known to have lived 32 years in the wild.

Bats get caught on barbed-wire fences, cacti, and thistles, and they occasionally fly into towers and other structures during storms. Baby bats usually die if they fall from the roost, and large colonies of bats can be killed if their caves are flooded. But the most significant causes of premature bat death are the activities of people.

Coachwhip Snake eating a Mexican Free-tailed Bat blown to the ground by a gust of wind at the cave entrance.

1. Fruit bat eats fruit of pioneer plant.

2. Deposits seeds in clearing.

3. Pioneer plants grow.

4. Rainforest restored.

PEOPLE AND BATS

ECOLOGICALLY SPEAKING, bats are very important. The quality of life for humans is directly related to a healthy global environment, and keeping that environment healthy requires bats. Bats are the primary predators of nocturnal insects around the world. Every night, they consume many millions of insects, including some that are crop pests.

Bats are often the most numerous mammals in the world's rainforests, where they are crucially important. For example, over 30 percent of all tree species on several tropical Pacific islands are pollinated or have their seeds dispersed by flying foxes. Bats also disperse the seeds of many pioneer plants that reforest cleared tropical lands. Other important plants pollinated by bats include the Baobab, a tree on African savannas, and the giant cacti of the Sonoran desert.

Durian flowers open at night and are pollinated by several species of bats. The trees then produce edible fruits *(right)*.

ECONOMICALLY SPEAKING, bats are important also. Hundreds of millions of dollars are generated each year by developing countries from plants that rely on bats for pollination and for seed dispersal. More than 300 economically important plant species, producing over 450 commercial products, are known to depend on bats in the Old World tropics alone. Just one of these products, the Durian fruit, contributes $120 million each year to the economy of Southeast Asia. Wild varieties of cultivated plants, such as agaves (from which tequila is produced), avocados, bananas, breadfruit, dates, figs, mangoes, and peaches, depend on bats. These wild plants are used for improving commercial varieties.

Insectivorous species play additional economic roles, such as the large-scale predation of insect pests. After an eradication campaign in Israel killed many insect-eating bats, harmful moth populations are said to have increased. The government had to use pesticides in an effort to control the moths. Certain bacteria from the droppings of an insect-eating bat have also been shown to aid in the treatment of chemical wastes and in the production of gasohol and detergents. Finally, the component of vampire bat saliva that prevents blood from clotting in a wound holds promise as an effective drug to prevent heart attacks in humans.

MOST BATS ARE CLEAN, HEALTHY, and unaggressive toward humans. There are, however, a few public health problems associated with bats, including the occurrence of rabies, although these problems are sometimes exaggerated by the press. Most of these problems are easily avoided by NEVER handling an unfamiliar bat and by not entering bat roosts without taking appropriate precautions.

However, less than 0.5 percent of all bats have rabies, a percentage that is no higher than that observed in most other mammals. Like other mammals, bats cannot carry the virus without becoming sick and dying. However, rabies does not spread extensively among bats and is only rarely transmitted by bats to other kinds of mammals, such as dogs or humans. Fewer people have died from bat rabies during the past 40 years than have died from dog bites or bee stings in a single year.

The risk of exposure to rabies could be reduced to near zero by vaccinating all dogs and cats and by avoiding contact with any wild animal that can easily be caught. If a person is bitten by a bat—or by *any* wild animal—medical advice should be sought immediately.

The rabies virus may also be transmitted to other mammals in caves where the air is saturated with vapor from millions of bats. But this is also extremely rare and can easily be avoided by not entering caves that house large bat colonies.

The bat droppings in some roosts contain the spores of a fungus that can cause the respiratory infection called histoplasmosis. Histoplasmosis can be lethal in humans. But more often infection results in flulike symptoms that usually disappear with medical treatment. This fungus is found in a variety of soils rich in organic matter, but people may pick up the infection by entering bat roosts where there is a lot of dust containing the spores.

49

BAT CONSERVATION

BATS ARE DISAPPEARING at an alarming rate. Humans are killing bats—intentionally or accidentally—around the world. Roosts are being destroyed or disturbed with regularity, foraging habitats are being lost, large numbers of bats are being consumed for food, and many have died from pesticide poisoning.

Nearly 40 percent of the bats in the United States are on the federal endangered species list or are candidates for it. Although protected by the Endangered Species Act, Indiana Bats have recently declined by approximately 55 percent. All but a few thousand Mexican Free-tailed Bats, out of a colony of 30 million, were apparently destroyed by vandals at an Arizona cave. The Carlsbad Caverns' population of this same species has declined from 8 million to less than a quarter of a million, probably caused by the use of agricultural pesticides. Almost all European bats are endangered or threatened. In Latin America and Israel, hundreds of cave roosts have been dynamited or poisoned in misguided efforts to control Common Vampire Bats and Egyptian Fruit Bats. At least four species of flying foxes have become extinct in recent years. Other species are severely threatened by habitat loss and unregulated hunting in the Indian Ocean on Pemba, the Comoro Islands, and Mauritius, and in the Pacific on the Marianas and Caroline islands. During a ten-year period, over 200,000 flying foxes were killed by commercial hunters from some of the small Pacific islands. The dead bats were then exported to Guam, where they were eaten.

In Australia and Africa, thousands, if not millions, of flying foxes have been killed because they were erroneously perceived as a threat to commercial fruit crops. Fruits eaten by these bats are usually too ripe to be harvested for commercial purposes.

YOU, TOO, CAN BECOME INVOLVED in trying to save bats. Unfortunately, so little is known about most bats that their population status can only be guessed. Many species may disappear before their decline is even recognized.

Action by members of organizations that promote the protection of bats is a vital part of any conservation effort. This is particularly true with bats because of the fact that so many people still harbor misconceptions about them. Individuals can directly influence bat conservation efforts by becoming active members of these organizations (see p. 150), which cannot survive without an influx of new members. Education is the most effective way to overcome personal biases. Amateur bat groups in Europe have developed many education programs and have helped gather data on the status of many species. Student bat clubs are now forming in the United States; you can become a member of one. Additionally, you can participate in bat programs at nature centers and in schools.

Chapman's Bare-backed Fruit Bat; to 7½ in. (195 mm); flying into a dark cave. Once found in the Philippines, this bat is now thought to be extinct.

CLASSIFICATION OF LIVING BATS

Individual bat species in this book are classified, or grouped, into related suborders, families, and genera. The order of the bat listings in this book is generally meant to reflect the evolutionary relationships among the various groups. Body lengths, provided in the captions, do not include the tail.

ORDER CHIROPTERA

SUBORDERS AND FAMILIES	# GENERA	# SPECIES
Suborder Megachiroptera		
Flying foxes and Old World fruit bats (Pteropodidae)	42	173
Suborder Microchiroptera		
Mouse-tailed bats (Rhinopomatidae)	1	3
Sheath-tailed bats (Emballonuridae)	12	48
Bumblebee bats (Craseonycteridae)	1	1
Slit-faced bats (Nycteridae)	1	13
Old World false vampire and Yellow-winged bats (Megadermatidae)	4	5
Horseshoe bats (Rhinolophidae)	1	69
Old World leaf-nosed bats (Hipposideridae)	9	63
Bulldog bats (Noctilionidae)	1	2
Short-tailed bats (Mystacinidae)	1	2
Mustached bats (Mormoopidae)	2	8
New World leaf-nosed and Vampire bats (Phyllostomidae)	48	148
Funnel-eared bats (Natalidae)	1	5
Smoky bats (Furipteridae)	2	2
Disk-winged bats (Thyropteridae)	1	2
Old World sucker-footed bats (Myzopodidae)	1	1
Plain-nosed bats (Vespertilionidae)	42	355
Free-tailed bats (Molossidae)	16	86
Totals	**186**	**986**

Straw-colored Flying Fox; to 7½ in. (190 mm);
male being greeted by harem females. Found in Africa.

FLYING FOXES AND OLD WORLD FRUIT BATS
Family Pteropodidae

Old World tropical landscapes are frequently populated by bat species in the family Pteropodidae, the only family in the suborder Megachiroptera. Pteropodid bats range from Africa eastward across India and Australia to the Cook Islands in the Pacific. They have foxlike faces with large eyes, simple ears, and, most typically, a big muzzle. Echolocation is generally unknown among these bats, which use sight for orientation and navigation. Some species may use more than 20 communication calls audible to people. Almost all members of the family have a claw on the second finger and the thumb, and in all but a few species the tail is short or absent. Weights vary from about half an ounce for the smallest nectar-feeding species to over 2 pounds for the large fruit-eating flying foxes. All species eat plant parts, especially fruit and nectar, and most produce only one pup per year.

FLYING FOXES These bats, which belong to the genus *Pteropus,* include the largest bats in the world. They are found on Madagascar and some islands in the Indian Ocean, in India and Southeast Asia, through Malaysia, the Philippines, Papua New Guinea, northern Australia, and eastward to the Cook Islands. The impressive Gigantic Flying Fox has a wingspan of about 6 feet and weighs over 2 pounds. The Large Philippine Flying Fox of the Philippine Islands weighs slightly more but has a smaller wingspan. Roosts of Large Flying Foxes once covered over 20 acres and contained over 100,000 bats. Today, roosts are much smaller, and many populations have disappeared altogether due to habitat loss and overhunting. Most (48) of the 57 species in this genus are found on islands; their populations are usually small and thus vulnerable to extinction.

The Samoan Flying Fox and the Tongan Flying Fox, with wingspans of up to 3 to 4 feet, are the flying foxes found

Comparison of *(left)* Gigantic Flying Fox; to 16 in. (400 mm) and Turkey Vulture.

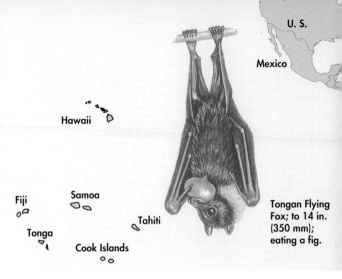

U. S.

Mexico

Hawaii

Fiji

Samoa

Tahiti

Tonga

Cook Islands

Tongan Flying Fox; to 14 in. (350 mm); eating a fig.

closest to the U.S. mainland. Both species are found on the Pacific islands of Samoa and Fiji. Tongan Flying Foxes also range from the Cook Islands, west of Tahiti, to islands off the coast of Papua New Guinea. Samoan Flying Foxes are distinguished by the amount of white on their head and by the light brown, rather than golden, fur on the chest and upper back. They usually roost alone and, unlike most other bats, forage during the day. Tongan Flying Foxes feed mostly at night and form large roosting camps.

Many bat populations in the Pacific have declined since 1970 as a result of commercial hunting. Bats are eaten by the native people of the Marianas Islands. Only about 500 of the endangered Marianas Flying Fox remain on the Pacific island of Guam, and the Lesser Marianas Flying Fox is already extinct. Since 30 to 40 percent of rainforest trees on the islands are pollinated by bats or their seeds are dispersed by bats, the loss of bat populations can have serious ecological and economic consequences.

Lesser Short-nosed Fruit Bats; to 5 in. (125 mm).

LESSER SHORT-NOSED FRUIT BAT This attractive brown bat, with white borders on its ears and wings, is a fairly small member of the family Pteropodidae. It ranges from Sri Lanka through southern Asia into Indonesia and the Philippines. Short-nosed fruit bats, including this species, are unique among members of their family in that they use plant leaves to form suitable roost sites. They also roost in palm trees, caves, and houses. These bats may produce a single pup twice a year. Some populations of short-nosed fruit bats are declining because the animals are caught and killed, with their body parts being sold and used as folk remedies for various ills.

QUEENSLAND TUBE-NOSED FRUIT BAT Tube-shaped nostrils angled to the side and yellow spots on their wings, forearms, and ears distinguish this bat, which has a dark stripe down its back. The bat's short, strong jaws are well adapted for crushing fruit to obtain the juice. Its mottled coloration may help camouflage the bat as it roosts under leaves or on the trunk of a tree. Females have one young each year. The exact function of the bat's unusual nostrils remains a mystery. This bat is found only in northern Australia.

Queensland Tube-nosed Fruit Bat; to 4¼ in. (110 mm).

Long-tongued Nectar Bat; to 2½ in. (65 mm); feeding on bottle brush.

LONG-TONGUED NECTAR BAT This tiny bat, which ranges from Thailand to the Solomon Islands and northern Australia, is grayish brown to reddish in color. It is one of the smallest megachiropteran bats and is smaller than many of the microchiropteran species. With a long tongue and muzzle, as well as incisors that are delicate and apart from the other teeth, Long-tongued Nectar Bats are specialized for eating nectar and pollen. They are important pollinators of Old World tropical forests, in palm and dipterocarp (a type of Asian tree) forests, and in mangrove and paperbark woodlands. Long-tongued Nectar Bats can apparently give birth during any season in parts of their range. Leaves on banana and hemp plants are used by these bats for roosting, which the Long-tongued Nectar Bat generally does alone.

QUEENSLAND BLOSSOM BAT Of the three species of blossom bats, this is the only one found in tropical northeast Australia. Also found in Papua New Guinea, Queensland Blossom Bats all have soft, long fur that is fawn to reddish in color. Bananas are among the many items in their diet. However, with a pointed muzzle and a tongue that has tiny brushlike projections down the center, they are especially good at obtaining nectar and pollen. A Queensland Blossom Bat will attack any other bat that flies close as it hovers in front of flowers. The roosting and reproductive behaviors of these bats are not well known. This species and the Long-tongued Nectar Bat are similiar in appearance, size, and color. They also appear to share similar tastes in food.

**Queensland Blossom Bat; to 2 in. (50 mm);
feeding on banksia blossom.**

MOUSE-TAILED BATS
Family Rhinopomatidae

Ancient pyramids and palaces, caves and cavelike openings, and houses are used by these primitive, insect-eating bats that look more like flying mice than like other bats. Mouse-tailed bats are easily identified by their long, thin tail, which is nearly as long as their head and body. The bats' soft fur is grayish to dark brown and is sparse or totally lacking on parts of the face and rump. They have a small, rounded nose leaf and ears joined at the base. The three species in this family belong to a single genus, *Rhinopoma,* and are usually found from the arid or semiarid, treeless regions of northern Africa through the Middle East to Thailand and Sumatra. Roosting colonies may contain many thousands of individuals.

GREATER MOUSE-TAILED BAT This bat, like the two other species in this family, gives birth to only one young per year each summer. In the fall the bat accumulates large fat deposits beneath the exposed lower back and base of its tail. This fat is used to provide energy in the winter, when temperatures are cool and insects are scarce. Mouse-tailed bats have a unique valve in their nostrils that can be closed at will, possibly to prevent the entry of sand and dust. Quite agile, these bats have maneuvered around the inside of Egyptian pyramids for at least 3,000 years. Their range is from Morocco to India and Sumatra.

Greater Mouse-tailed Bats; to 3 in. (75 mm).

Pacific Sheath-tailed Bat; to 1¾ in. (45 mm); on small island. Found on the Marianas and Caroline islands. This and a closely related species are disappearing in many parts of their range.

SHEATH-TAILED BATS
Family Emballonuridae

Bats in this family have a tail that begins near the middle of the interfemoral (tail) membrane. This tail is enclosed in a special sheath made of skin that connects with the tail membrane. During flight, when the hind legs are stretched out, the membrane is forced to slide over the tail, thereby increasing the total amount of flight surface area. Many species also have sacs, or pockets, on the front edge of their wings. These sacs, as well as the throat sacs of several species, are most highly developed in males and produce a musky secretion that is thought to be used to mark territories. Most species are small to medium in size. Color varies from grayish or dark brown to black and even white in a few species. Roosting bats in this family frequently hold their folded wings at an angle away from the body.

PROBOSCIS BAT This bat is often seen by travelers on tropical rivers in the Americas from southern Mexico to Bolivia and Brazil. Proboscis Bats may be mistaken for moths when disturbed and flushed from roosts. Small groups of about a dozen individuals roost on branches, fallen trees, and rocks near water. The brown fur flecked with white, wavy lines on their back, and the tufts of gray hair on their forearms make these bats difficult to see even in daylight. As its name suggests, the Proboscis Bat has a long, fleshy nose, or proboscis. It feeds on very small insects captured over water. Proboscis Bats roost and hunt in family groups. Roosts and feeding territories are defended by dominant males. Females typically give birth to a single pup but can have two young per year.

Proboscis Bats; to 1½ in.
(40 mm); roosting on a log.

Northern Ghost Bat; to 3 in. (75 mm).

NORTHERN GHOST BAT This species has long, silky fur that is white to pale gray in color. It is quite unique in the bat world. It differs from the Honduran White Bat (not shown) in being larger and lacking a nose leaf. Instead of wing sacs, the Northern Ghost Bat has sacs attached to its interfemoral membrane, near the tail.

Northern Ghost Bats range from southern Mexico south to the Amazon Basin of Peru and Brazil. They are found in most tropical habitats, including rainy and dry forests, and in cultivated areas such as coconut plantations. They are insectivores and usually forage high above open areas or above the forest canopy. They are also attracted to insects around streetlights. Most of these bats roost alone under palm fronds except during mating, when small harems may be formed. A single pup is born during the summer.

TOMB BATS As the name implies, tomb bats are numerous in some of the larger tombs and pyramids of Africa. Some species also roost in rock crevices, caves, and trees. The combined range of the 20 insectivorous species of tomb bat includes Africa and Madagascar through southern Asia and Australia to the Solomon and Philippine islands in the Pacific. The fur of these bats is gray, brown, or reddish brown above and paler below. Tomb bats can be identified in flight by their long, narrow, whitish wings, which are somewhat translucent. Males of most species have well-developed throat sacs in addition to pockets on the wings. Several species produce echolocation calls that humans can hear. Most species give birth to only one pup a year. Tomb bats accumulate fat during the fall and may enter periods of decreased activity when insects become scarce.

Egyptian Tomb Bat; to 4 in. (100 mm). Found in the savanna regions of Africa and Madagascar.

BUMBLEBEE BATS
Family Craseonycteridae

This family is unusual in that it includes only one genus with one species. The bumblebee bat was not identified until 1973 and is the most recently discovered family of bats. In some ways, this family is similar to mouse-tailed bats and sheath-tailed bats. Lack of a tail and the presence of relatively long, broad wings that appear to be used for hovering in the air distinguish this bat.

BUMBLEBEE BAT This is the smallest bat known and may be the smallest mammal. Bumblebee Bats weigh about the same as a penny. They are sometimes called Old World Hog-nosed Bats due to to their piglike nose. They also have relatively large ears.

Bumblebee Bats have an extremely limited geographical distribution. They have been located exclusively in or near certain limestone caves in a small region of west-central Thailand. Individuals have been observed flying over the tops of bamboo thickets and stands of teak trees. They use echolocation to capture a variety of small insects in flight and may swoop down to pick off insects from foliage. Females produce single pups in May.

The small populations of these tiny bats are threatened by a variety of factors, especially the loss of forests from commercial logging and the disturbance of roosts by cave visitors. The government of Thailand has recently taken action to protect this unique little mammal.

Bumblebee Bats (actual size); to 1¼ in. (32 mm).

SLIT-FACED BATS
Family Nycteridae

A deep groove bordered by flaps of skin along the top of the muzzle, plus a hollow pit at the base of the groove, distinguishes these bats as a family. Their long tail, with its T-shaped tip, is unique among bats. The tail is completely enclosed within the interfemoral membrane. Slit-faced bats have large ears that appear separated but are actually joined at the base by a low flap. The color of the long, loose fur is rich brown or rust to pale brown or gray. The 12 species belonging to the single genus of this family can be found in rainforests and in more arid regions, from Africa and Madagascar to Saudi Arabia and Israel. One species ranges from southern Burma to Indonesia. Caves are the preferred roosts, but slit-faced bats are sometime found in tunnels, buildings, trees, and even porcupine and aardvark burrows.

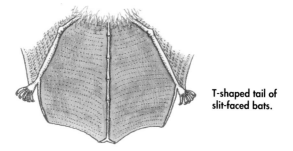

T-shaped tail of slit-faced bats.

LARGE SLIT-FACED BAT Nycterid bats consume an impressive variety of prey, including moths and other insects, spiders, and scorpions. Some species, including the Large Slit-faced Bat, also eat fish, frogs, birds, and even other bats. Large Slit-faced Bats are found in Africa.

Large Slit-faced Bat;
to 3 in. (75 mm);
eating a horseshoe bat.

Yellow-winged Bat; to 3 in. (75 mm); found in Central Africa.

OLD WORLD FALSE VAMPIRE AND
YELLOW-WINGED BATS
Family Megadermatidae

Members of this family are large. They have big eyes and large ears. The upper incisors are missing, and the tail is either absent or very short. They have a fleshy, long, erect nose leaf that is not as complex as in horseshoe or Old World leaf-nosed bats. Although they eat mainly small vertebrates, Old World false vampire bats were given their name because it was once thought that they fed on blood. The range of the family is from tropical Africa and India to the Philippines and Australia.

AUSTRALIAN GHOST BAT This is one of the largest microchiropteran bats in the world. The name refers to the whitish head, ears, neck, chest, and wing membranes. This endangered species is found only in the northern half of Australia, where it lives in arid and rainforest habitats. It is the only carnivorous bat in Australia, eating mostly small birds, reptiles, and small mammals, including other bats. Australian Ghost Bats use vision, echolocation, and the variety of noises made by prey to locate food. Mating probably takes place in July or August; a single pup is born between September and November. A mother bat brings food to her pup, and she later teaches it how to hunt. Distribution is patchy, and many populations are dangerously low. Australian Ghost Bats are particularly sensitive to disturbances of their cave roosts, many of which have been destroyed by mining activities.

Australian Ghost Bat; to 4½ in. (115 mm); sharing a frog with its young.

71

HORSESHOE BATS
Family Rhinolophidae

These bats have an elaborate and distinctive nose leaf. The lower part is horseshoe-shaped, the middle section varies in shape, and the top part tapers to a point. The ears are large, with a fold on the outside edge, and they lack a tragus, or flap inside the ear. Horseshoe bats have short tails, and their color varies from reddish brown to black. Although there is only one genus, there are nearly 70 species, found in Europe, Africa, Japan, Southeast Asia, Papua New Guinea, and Australia. Unlike most echolocating bats, horseshoe bats emit ultrasonic calls through the nostrils rather than through the mouth. The nose leaf directs the echolocating calls forward. Many species roost in caves or hollow trees, where colony members wrap their wings around their bodies and keep a little distance between them and their neighbors. Northern populations usually hibernate in caves. All species are insectivorous. Some scoop insects from the ground.

Front and profile of Greater Horseshoe Bat; from 2½ to 5 in. (65 to125 mm).

Greater Horseshoe Bat female *(right)*, with young attached to false nipple, about to land on cave ceiling near roostmates.

GREATER HORSESHOE BAT This species, one of the largest rhinolophids, occupies a variety of habitats, from pastures to forests. Greater Horseshoe Bats mate in the fall. They bear a single young or twins in the spring, usually in maternity colonies located in buildings. Females do not usually mature sexually until they are 3 to 7 years old—very late for a bat. They seem to stop reproducing at around 10 years of age. European populations are endangered, particularly in Britain, where only about 1 percent, or 3,000 bats, remain from the original population as estimated at around 1900. Destruction or disturbance of roosts seems to have caused these drastic declines. These bats are found from England to India and Japan.

OLD WORLD LEAF-NOSED BATS
Family Hipposideridae

The front part of the ornate nose leaf of Old World leaf-nosed bats is shaped like that of the horseshoe bats. The middle section differs in that it is cushionlike, without a raised point. The back part of the leaf is rounded rather than lance-shaped and may be divided into cells or points. Each toe of an Old World leaf-nosed bat has two bones instead of three, as in horseshoe bats. Old World leaf-nosed bats range from Africa through India and Southeast Asia to northern Australia. They catch most of their insect prey on the wing. These bats roost in caves, hollow trees, and buildings, which they frequently share with horseshoe bats. One species in Africa has been observed using porcupine burrows.

Tail-less Leaf-nosed Bat *(upper left)*; **African Trident-nosed Bat** *(upper right)*; **Commerson's Leaf-nosed Bat** *(below)*.

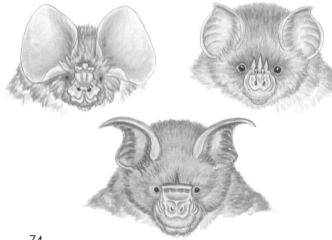

Diadem Leaf-nosed Bat; to 3 in. (75 mm); eating a flying ant.

DIADEM LEAF-NOSED BAT A nose leaf with three or four ridges on the sides, as well as white to pale yellow spots on its dark brown back and sides, identifies this large bat. The Diadem Leaf-nosed Bat roosts in caves, trees, culverts, and buildings. It may hunt for insects on the fly or wait in ambush at a perch. If a swarm of insects is located, a bat may catch many insects and store them in cheek pouches until returning to its perch. A single pup is born between November and January. This bat's range is Southeast Asia and the Philippine Islands to the Solomon Islands and northern Australia.

BULLDOG BATS
Family Noctilionidae

The enlarged, drooping lips—which are split, exposing the lower canines—and the absence of a nose leaf give the two bat species in this family a bulldog look. Bulldog bats have very short fur that is rust-colored or reddish to dull brown or gray. There is also a faint stripe down the middle of the back. The ears are long, pointed, and slant forward. Feet and toes are large, with long, curved, sharp claws. These bats are found in lowland lake, river, and coastal habitats, where both species can be recognized by their large size, long and narrow wings, and the behavior of dipping their feet as they skim over water. They start flying at dusk. Small flocks often forage together in circular formation. Lesser Bulldog Bats feed almost exclusively on insects, especially water beetles. Both Lesser and Greater

Bulldog Bats roost in colonies in hollow trees and shallow caves, but only the Lesser Bulldog Bat also roosts in buildings. Both species are good swimmers and can take off from water.

GREATER BULLDOG OR FISHING BAT This species, found in the Antilles and along the Pacific coast of northern Uruguay, is adapted for locating and capturing fish. After using echolocation to detect an exposed fin or to pick up a disturbance made by a fish near the water's surface, the Greater Bulldog Bat dips down, lifts its tail, and drags its claws through the water until the fish is speared. The prey is then transferred to cheek pouches until the bat reaches a perch, where it eats the fish. Greater Bulldog Bats also capture insects, and they roost in caves. Mating in populations north of the equator takes place in November or December; a single pup is born in late spring.

Greater Bulldog Bat; to 5 in. (125 mm); sequence shows it catching a fish.

SHORT-TAILED BATS
Family Mystacinidae

Although found only in New Zealand, the two species in this family are most closely related to tropical American bats in the families Noctilionidae, Mormoopidae, and Phyllostomidae. This is not surprising, given that much of the flora and fauna of New Zealand first evolved in South America.

Short-tailed bats are uniquely adapted for a terrestrial existence. The wing membranes are thick and leathery along the forearms and sides of the body. The wings can be folded up and partially tucked into a body pouch. The claws of the thumb and feet each have a small, additional claw near their base. Short-tailed bats have short, stout legs with large, broad feet. Their soles are soft, and their fur is thicker than that of any other bat. The tongue has a small, brushlike patch at its tip. One species may already be extinct.

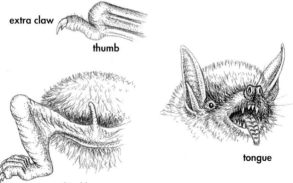

extra claw

thumb

hind leg

tongue

The thumbs and feet of Lesser Short-tailed Bats have small, additional claws at the base. The tongue is deeply ridged, with a small, brushlike tip.

Lesser Short-tailed Bat; to 2½ in. (60 mm); about to enter its burrow. Shown actual size.

LESSER SHORT-TAILED BAT This New Zealand bat is highly terrestrial, as are many other Pacific island vertebrates. It is also ecologically very versatile, feeding on aerial and ground insects, fruit, pollen, nectar, and carrion. Colonies roost in hollow trees, caves, seabird burrows, and holes in cliffs. They also dig cavities in fallen logs for roosts. These bats are very agile on the ground, where they use their folded wings for digging, walking, running, and climbing. They seldom fly more than a few miles. Lesser Short-tailed Bats are grayish brown to brown. They do not hibernate. Males form leks, where they congregate to attract females. A single pup is born in December or January.

Davy's Naked-backed Mustached Bats; to
2 in (50 mm). Found from Mexico south to
Brazil, Lesser Antilles, and Trinidad.

MUSTACHED BATS
Family Mormoopidae

The faces of these odd-looking bats may have evolved
because of the special way they capture insects. Both the
large lips that flare outward and a mustache of stiff hairs
give the open-mouthed bat a distinctive look. The funnel-
shaped mouth may help to direct echolocation calls as well
as provide a large trap for capturing insects in flight. The
lower edge of the ear is connected to the mouth by a fleshy
ridge. The tail begins about halfway down the interfemoral
membrane. The wing membranes are attached to the body
relatively high on the sides. The wings are attached to the
center of the back in two species, both naked-backed
mustached bats. Mustached bats are fast flyers with long,
narrow wings. They range from southern Arizona and
Texas through Central America, the Caribbean islands, and
parts of South America. Their local distribution may be
limited by the availability of warm, humid caves.

LEAF-CHINNED BAT The unusual and distinctive appearance of this bat's head is due to the two curved skin flaps on the chin, the mustache of stiff hairs, and the complex ears that are connected to each other and to the mouth, forming a funnel. Color of the long fur varies from bright cinnamon to brown, yellowish white, or brownish gray. Leaf-chinned Bats are usually found in dry habitats but have been spotted in rainforests. They fly swiftly as they forage for insects close to the ground or above water. This species, ranging from southern Texas and Arizona south to northern South America and northwestern Peru, typically roosts in large, warm caves, where numbers may range from a few individuals to about half a million bats. They may roost in the same cave with—but apart from—other species, such as the Mexican Free-tailed Bat. Unlike most other cave bats, Leaf-chinned Bats maintain a distance from each other in the roost. A single pup is born in late spring or summer.

Face of Leaf-chinned Bat (larger than life-size);
to 2½ in. (65 mm);
showing unique facial ornaments.

NEW WORLD LEAF-NOSED AND VAMPIRE BATS
Family Phyllostomidae

Although this is the most diverse family of bats with respect to feeding habits, Phyllostomidae does not contain as many species as do the families Pteropodidae and Vespertilionidae. New World leaf-nosed bats consume fruits, nectar, pollen, insects, other vertebrates, and even blood. They occupy a variety of habitats from deserts to high-elevation forests, and are found from the southern United States to southern South America. They are the only group in the New World with a nose leaf. This structure is relatively simple and spear-shaped in most species and extremely small in vampires. The nose leaf may help direct the echolocation calls emitted through the nostrils. Calls are also emitted through the mouth. All species have a tragus (ear flap) and rather simple ears. Color varies from black and brown to orange-brown and white. Many species have white or yellow markings on the face, ears, back, or wings. There are several very small species in this family, as well as the Neotropical False Vampire Bat, which is the largest of all the microchiropteran bats. The 150 or so distinct species are divided up into six different subgroups, or subfamilies: Phyllostominae, Glossophaginae, Carolliinae, Stenodermatinae, Brachyphyllinae, and Desmodontinae.

SPEAR-NOSED BATS
(Subfamily Phyllostominae)

All of these bats have a nose leaf, with a horseshoe-shaped base and a large, erect, lance-shaped part. Spear-nosed bats also have a tail with all but the extreme tip enclosed in the tail membrane. They usually have large or very large ears. This subfamily includes all of the

Brazilian Little Big-eared Bat; to 1¾ inches (45 mm); eating a moth.

carnivorous members of the family as well as those that eat mostly insects. Many other species in the subfamily eat both fruits and insects.

LITTLE BIG-EARED BATS The ten species of this genus range from southern Mexico to Brazil. As their name implies, these are small bats with large ears. The lower jaw has smooth pads of skin on each side that come together to form a V-shaped chin. The tail is short, and the tail membrane is the same length or shorter than the legs. The long fur is brown to chestnut or orange on the upper parts, lighter below. In some species the hairy ears may be joined at the base. Little big-eared bats use feeding perches, where they consume large insects, such as cockroaches and moths, and fruits. These alert bats roost in small groups near the ground in damp, secretive sites, such as inside or beneath hollow logs, among tree roots, in small caves, on banks beside streams, and in culverts.

CALIFORNIA LEAF-NOSED BAT This is the only leaf-nosed bat that is a permanent resident of the United States. It is found in desert habitats from southern California and Arizona into northwestern Mexico. The bat's long tail extends slightly beyond the end of the tail membrane. Its nose leaf distinguishes this species from the other big-eared bats in the region.

California Leaf-nosed Bats forage while flying slowly just above the ground. The soft, fluttering sounds of their flight are often used to identify these bats on the wing. They apparently use vision, echolocation, and the noises made by prey to locate insects, which are plucked from the ground or from low-lying vegetation. Prey includes grasshoppers, cicadas, moths, caterpillars, and beetles. Mating takes place in the fall, followed by a period of delayed embryonic development that may last for five months. A single young or twins are born from May to July.

California Leaf-nosed Bat; to 2½ in. (65 mm);
hovering above a cricket.

84

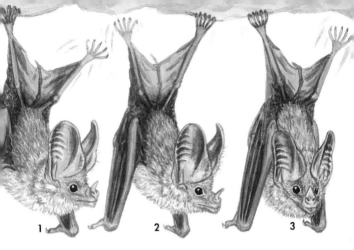

1   2   3

California Leaf-nosed Bat moving across cave ceiling.

Abandoned mines and small caves are the preferred roosts of California Leaf-nosed Bats. These structures protect the bats from extreme heat during the summer and from cold temperatures in winter. The bats do not hibernate or migrate and will die if their body temperature falls below 79° F. Bats hang from the horizontal surfaces of their roosts in groups of several to 100 or more individuals, separated from one another by a few inches. Unlike most other bats, California Leaf-nosed Bats are unable to crawl while hanging from the roost. They move about the roost by swinging one foot at a time in a stepwise fashion. Populations of the California Leaf-nosed Bat have recently declined because of human disturbances of their roosts. These bats are now considered candidates for the U.S. endangered species list.

SPEAR-NOSED BATS This genus is comprised of large bats with a rounded head, a powerful muzzle, and a V-shaped row of bumps on the chin. The triangular ears are widely separated and the nose leaf is simple and broad. The fur is short and velvety and usually black but can be grayish brown or reddish, sometimes with a silvery sheen. Spear-nosed bats generally occupy forest habitats. Females give birth twice during the rainy season. They most often roost in large colonies in caves or tree hollows, where harems of up to 25 females and a male are formed. Fruits, pollen, nectar, and insects are the most commonly eaten foods. The largest species, the Greater Spear-nosed Bat, whose range is from Honduras to Brazil and Peru, also occasionally eats other vertebrates.

Greater Spear-nosed Bats; to 5 in. (125 mm);
male defending harem females.

Neotropical False Vampire Bat; to 6½ in. (165 mm).

NEOTROPICAL FALSE VAMPIRE BAT This huge flying carnivore, found from southern Mexico to northern Bolivia, is distinguished by its size. With a wingspan as great as that of some flying foxes, the Neotropical False Vampire Bat is the largest of the microchiropteran bats. The head is also distinctive, with very long ears, large eyes, and a long muzzle. Like most other predators, these bats are uncommon throughout their range. The Neotropical False Vampire Bat searches for birds, other bats, and rodents in forests, over swamps, and along forest edges. Large food items may be shared by family members, which roost in hollow trees. Mated bats may remain together for their entire lives.

LONG-TONGUED BATS
(Subfamily Glossophaginae)

These bats are small and delicate, with a long, narrow muzzle. The long, slender tongue can be extended nearly one-third the length of the bat's body and usually has hairlike projections on the tip. The lower incisors are small or absent, and the lower lip has a V-shaped notch or grooved section. Fine whiskers sprout from the chin and around the short, triangular-shaped nose leaf. Many species consume insects in addition to feeding on nectar. Although these bats are very important as pollinators of tropical plants, little is known about most of the species in this subfamily. One genus, *Monophyllus,* is restricted to the Bahamas and the Antilles.

Peruvian Spear-nosed Long-tongued Bat; to 2¾ in. (62 mm); chasing another bat from roost site. Found in Ecuador and Peru.

Mexican Long-tongued Bat; to 3 in. (75 mm). In southeast Arizona, these bats can sometimes be seen visiting hummingbird feeders.

MEXICAN LONG-TONGUED BAT A short tail and a more slender, longer muzzle separate this species from the Greater and Lesser Long-nosed Bats. Its range is from the southern boundaries of states bordering Mexico to just north of Nicaragua. Mexican Long-tongued Bats occupy a variety of habitats, from arid terrains to mixed oak-conifer forests. Their diet includes pollen, nectar, and the fruit of cacti, *Agave,* and other night-blooming plants. They may also consume insects. Their migratory pattern is similar to that of Greater and Lesser Long-nosed Bats. Mexican Long-tongued Bats roost in small caves or near the entrances to large caves. Females give birth to a single pup in early spring. Roosting bats are easily disturbed, and the Mexican Long-tongued Bat is now considered a candidate for the U.S. endangered species list.

COMMON LONG-TONGUED BAT This bat has fur that is whitish at the base and dark gray or brown at the tip. The horseshoe part of the nose leaf is attached to the muzzle, which is not very long. The V on the chin contains a row of bumps. Common Long-tongued Bats have a very short tail. They are found from Mexico to Argentina, where they consume nectar, pollen, fruits, and insects in a variety of habitats and form colonies of up to several hundred individuals in caves and cavelike structures, hollow trees, and buildings. This species and the Common Short-tailed Fruit Bat (p. 93) frequently roost together. Young may be born in any season. The four or five other species in this genus, *Glossophaga*, are similar in appearance.

GREATER AND LESSER LONG-NOSED BATS These two closely related species were recently added to the list of endangered species in the United States. Both are quite large for glossophagine bats and have fairly long, sloping muzzles with small, triangular nose leaves. Characterized by a very narrow tail membrane, these bats do not have an obvious tail. Lesser Long-nosed Bats are gray to reddish brown and have a wingspan of about 15 inches. Greater Long-nosed Bats are sooty brown and are about 10 percent larger. The former species is found in dry desert habitats from southern Arizona and New Mexico southward to Baja California and Oaxaca. The latter is found in cooler, higher habitats from the Big Bend region of Texas south to Guatemala.

Both species feed on the nectar, pollen, and fruit of desert plants, including species of *Agave* and giant cacti such as the Saguaro, Cardon, and Organ Pipe. Small groups travel long distances each night to find open flowers. Their wings make distinctive whooshing sounds as they circle a plant, probably using odor to determine

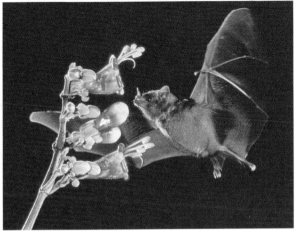

Common Long-tongued Bat; to 3 in. (75 mm); pollinating a Century Plant.

nectar levels. A bat will hover and then plunge its head into a flower to lap up the nectar, becoming covered with pollen in the process. Pollination takes place when the bat moves to another plant. Mexican populations migrate north to reproduce during the summer, following the seasonal blossoming of various desert plants. Colonies roost in caves.

Many of the plants pollinated by these bats are crucial to desert ecosystems and are economically important. Unfortunately, habitat destruction and cave disturbances have caused populations of Greater and Lesser Long-nosed Bats to decline dramatically. Conservation efforts are badly needed to ensure the survival of these bats and the desert ecosystems they help to maintain.

HAIRY-LEGGED LONG-TONGUED BATS These bats have short grayish to blackish brown fur, a long muzzle, and a tail that is extremely tiny or absent. The tail membrane is reduced to a hairy, very narrow band along the legs. Hairy-legged long-tongued bats range from tropical Mexico to Bolivia and southeastern Brazil. The four species differ in color and size, with total length varying from 2 inches to slightly more than 3 inches. These bats feed on nectar, pollen, fruits, and insects. Their preferred roosts are caves, mines, and culverts, where colonies may number anywhere from a few to several hundred individuals. Although common in the Andes up to about 12,000 feet, these bats are rare or absent in areas lacking numerous caves, such as the Amazon Basin.

Geoffroy's Hairy-legged
Long-tongued Bat;
to 3 in. (75 mm);
feeding on nectar.

SHORT-TAILED FRUIT BATS
(Subfamily Carolliinae)

Members of this group are among the most abundant species in many lowland rainforest habitats in the New World tropics. They have medium-sized muzzles, nose leaves, and ears. Tails are short or absent, and tail membranes are moderate to small. Their teeth distinguish them from the other subfamilies.

Common Short-tailed Fruit Bat; to 3 in. (75 mm); carrying a piper fruit in its mouth.

SHORT-TAILED FRUIT BATS The bats in this genus are common and have soft fur that has three or four bands ranging in color from gray to brown. The muzzle has a large, spear-shaped nose leaf and a horseshoe-shaped flap. The chin contains a V-shaped line of small bumps and a large central bump. The medium-sized ears are triangular. Unlike the tailless little fruit bats, the other genus in this subfamily, a short tail extends above the tail membrane. The diet consists of fruits, insects, and sometimes nectar. These bats are found in lowland habitats, from southern Mexico to Bolivia and southern Brazil. They often carry the seeds of colonizing plants into open areas and thus are crucial to the reforestation of tropical lands that have been cleared. Bachelors and small harems roost in foliage, caves, and tree hollows, often with other species.

Common Yellow-eared Bat; to 2 in. (50 mm); under a philodendron leaf. This New World fruit bat is found from Mexico to northern South America around the Amazon Basin to southern Brazil.

NEW WORLD FRUIT BATS
(Subfamily Stenodermatinae)

All members of this large subfamily eat fruit but some may eat nectar or insects if fruit is scarce. New World fruit bats are the main seed dispersers for many rainforest plants. Like short-tailed fruit bats, they also carry seeds of pioneer plants into cleared areas and are therefore essential to the recovery of deforested land. The bat's muzzle is short and broad, and its teeth are flattened, an adaptation to its fruit diet. A fleshy nose leaf, no tail, and a short or absent tail membrane further distinguish these bats. Many species have whitish stripes on the face and back that may act as camouflage while they are roosting in foliage. Most species give birth twice yearly, during the rainy season.

YELLOW-SHOULDERED FRUIT BATS The shoulder region of these bats, especially the males, is usually stained yellow or orange to deep red by glandular secretions. Although these bats are common, little is known about them—including the function of their yellow shoulders or where they roost. They have pale brown to yellow-brown or gray fur, which is dark at the tip and base and beige or silver in the middle. Head and body length for the 12 species of yellow-shouldered fruit bats varies from 2 to 4 inches. A short, broad nose leaf, no tail, and an extremely narrow tail membrane are found on all of these bats, which appear to be most numerous in open areas of forests and near streams from lowlands to forests at high elevations. A few small colonies have been observed roosting in small caves in the Andes.

Common Yellow-shouldered Fruit Bat; to 2½ in. (65 mm); found from northern Mexico south to Argentina and Uruguay.

NEW WORLD FRUIT BATS The 21 species belonging to this genus, *Artibeus,* are divided into two separate groups of small species and slightly larger species. All species are black to gray or grayish brown with no stripe down the back. The head is large and powerful, the muzzle is short and broad with a fleshy and broad nose leaf, and the upper teeth are in a horseshoe-shaped row. Lower incisors have two lobes. A V-shaped row of bumps on the chin, a muscular body, and a short, U-shaped tail membrane are typical features of these bats. Many, but not all, species of New World fruit bat have whitish stripes on the face. New World fruit bats are found from central Mexico to northern Argentina and Paraguay. They occur in a variety of habitats, from highlands to lowland rainforests.

Head and body length in the nine larger species varies from almost 3 inches to over 5 inches. The largest of these, especially Big Fruit Bats and Jamaican Fruit Bats, are among the most common bats in lowland rainforests from Mexico and the Antilles to northern Argentina. New World fruit bats are especially fond of figs and other fruits but may also eat insects. The 12 species in the other, smaller group range in size from less than 2 inches to almost 3 inches. These

Heads (actual size) of the largest (Big Fruit Bat) and smallest (Pygmy Fruit Bat) of the New World fruit bats.

**Jamaican Fruit Bats; to 3½ in. (85 mm);
male returning to harem females in a hollow tree roost.**

bats feed on smaller fruits and insects. Both groups are
extremely important seed dispersers and can frequently be
seen at dusk flying down forest trails. They help maintain
existing rainforests and help reestablish forests on cleared
tropical lands.

New World fruit bats roost in tree hollows, among
leaves, under palms, and in caves. They will occasionally
even roost in buildings. Some species make tents by cutting
into the veins of large leaves, causing the leaves to droop.
Several species form harems in tree hollows and caves.

Small harem of Peter's Tent-making Bats; to 2½ in. (65 mm); seen here roosting under a palm leaf modified to form a tent, as shown in inset.

PETER'S TENT-MAKING BAT This interesting bat bites a line down each side of the midrib of large leaves such as palm leaves or the leaves on banana plants. This causes the sides to droop down and form a tent, where harems and family units roost. White stripes on the dark gray to dark brown head and back may camouflage this bat at the roost. The nose leaf has a large flap around the base, which, like the ears, is outlined in white. Peter's Tent-making Bat has a blackish tongue and lips and a hairless, deeply notched tail membrane. It is found in a variety of lowland habitats, from southern Mexico to Bolivia and southeastern Brazil.

WHITE-LINED FRUIT BATS These bats are similar to Peter's Tent-making Bats but have pink tongues, tail membranes fringed with hair, distinct dental features, and fainter markings, except for the white stripe down the middle of the back. The ten species of muscular bats have head and body lengths that range from 2 to 4 inches. They are found in mature and second-growth forests, arid scrublands, and high-elevation forests, from southern Mexico to northern Argentina and Uruguay. Small groups, probably harems and maternity colonies, roost in caves or under leaves. Little is known about most species, although they appear to be fairly common.

Montane White-lined Fruit Bat; to 2¾ in. (70 mm); shown here eating a fig.

WRINKLE-FACED BAT This bat, which inhabits tropical Mexico to northern Venezuela and the island of Trinidad, is unquestionably unique in appearance. The head is rounded. The hairless face is marked by many deep wrinkles, folds, and channels. The chin has a loose fold of skin that can be pulled up over the wide mouth and eyes. Interestingly, the membrane area covering the bulging eyes is translucent, which allows the bat to perceive objects while resting in the foliage of trees. Wrinkle-faced Bats also have distinctive wing membranes between the fourth and fifth fingers, where alternating translucent and opaque bands form a zebralike pattern. These bats extract juice from fruits in a wide range of forested habitats, including city parks. Their flight is erratic and resembles that of beetles.

Wrinkle-faced Bat;
to 3 in. (75 mm).

Lesser Antillean Fruit-eating Bat; to 3½ in. (92 mm); chasing Jamaican Fruit Bats from the roost.

FLOWER BATS
(Subfamily Brachyphyllinae)

With a long muzzle and tongue, these bats resemble members of the subfamily Glossophaginae (see p. 88). They differ in dental features and by having a small or only partly developed nose leaf. Flower bats are confined to several islands of the Caribbean.

ANTILLEAN FRUIT-EATING BATS The four species of these stocky bats are grayish to yellowish white above, with the hair tipped brown on the shoulder region, neck, and head. The muzzle is shorter than in the Common Long-tongued Bat and more conelike in shape. A nose leaf and tail are virtually absent. Antillean fruit-eating bats have shorter wings than other members of this family. These bats are restricted to the Greater and Lesser Antilles and the Bahamas, where they feed on nectar, pollen, fruits, and insects in dry and wet forest habitats. Roosts include deep caves, crevices, wells, and buildings. The Lesser Antillean Fruit-eating Bat lives only in caves in Puerto Rico and Barbados.

VAMPIRE BATS
(Subfamily Desmodontinae)

These are the only true vampire bats in the world. Until recently they were assigned to a separate family based largely on their feeding habits. Vampire bats have razor-sharp upper incisors and bladelike canines. The nose leaf is diminished and consists of small folds above the nose. The tail is absent. The thumbs are long and thick and are used with the dexterous forearms to walk, hop, run, and climb around a roost or quietly approach prey.

Vampires use their extremely sharp teeth to make a small V-shaped wound in the skin of their prey. They lick, rather than suck, the blood from the cut. An anticoagulant in the bat's saliva prevents the victim's blood from forming clots. Common Vampire Bats can feed on a wide variety of mammals, including humans, but domestic livestock has become their main prey in recent times. Because they have access to abundant food sources year-round, they can breed throughout the year. White-winged Vampire Bats and Hairy-legged Vampire Bats prefer birds. All three species roost in caves, hollow trees, tunnels, and buildings.

Common Vampire Bats can transmit rabies to livestock and humans, and the wounds they inflict are subject to screwfly infestations. However, their populations can be controlled without harming beneficial bat species in areas where vampires have become too numerous. It is very important that people do not carelessly eradicate all bat populations in an effort to control vampires, because most bats are economically and ecologically important.

The Common Vampire, found from Mexico to Chile and Argentina, is sleek and muscular; its fur is short, often with a silvery sheen. Small groups of females form stable roosting groups and share foraging territories. They also share regurgitated blood with their pups and adult roost mates.

Common Vampire Bat; to 3½ in. (90 mm);
sharing a blood meal with a sick roostmate.

103

Mexican Funnel-eared Bats; to 2½ in. (60 mm).

FUNNEL-EARED BATS
Family Natalidae

The alert manner, delicate body, and bright colors prompt some people to refer to these bats as "cute." Funnel-eared bats are distinguished by large, funnel-shaped ears that face forward, a long tail (equal in length to the head and body) enclosed in a pointed tail membrane, and legs as long as the tail. These bats have a high, domed crown that rises abruptly above the muzzle.

Adult males have a large glandular structure, called the natalid organ, at the base of the muzzle. Its function is unknown. The soft, long fur varies in color from beige, yellowish brown, or gold to bright rust. The four species of funnel-eared bats are restricted to the New World lowland

tropics, from Mexico and the Caribbean islands to southern Brazil. The fluttering, mothlike flight of these insectivores is distinctive. Funnel-eared bats typically roost in dark, humid caves, where they may number from a few dozen to many thousands. Little is known about the reproductive cycle or social behavior of these bats. This small family is probably related to the smoky bats (Furipteridae) and disk-winged bats (Thyropteridae).

MEXICAN FUNNEL-EARED BAT This is the largest funnel-eared bat. Funnel-eared bats breed during the dry season in southern Mexico, and females bear one pup. They typically roost in warm, humid caves, where up to nine other species may also be found. The Mexican Funnel-eared Bat was once also found in Cuba but is now extinct there.

SMOKY BATS
Family Furipteridae

The head and ears of these bats resemble those of funnel-eared bats. Smoky bats, also called thumbless bats, have their thumb and tiny claw partially enclosed within the wing membrane, so that they appear thumbless. The tail is also relatively short, only about half the length of the tail membrane. The dense fur is coarse and is gray to grayish brown in color. Smoky bats are among the smallest bats known, ranging from a little less than 1½ inches to slightly more than 2 inches in length. Both species in this family feed on insects. Unfortunately, not much is known about these intriguing little bats. Eastern Smoky Bats inhabit rainforest areas from Costa Rica and Trinidad into Peru and Brazil.

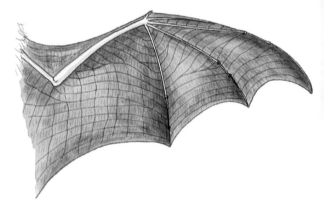

Thumbless wing of smoky bats.

Western Smoky Bats (roosting bats are actual size);
to 2 in. (50 mm).

WESTERN SMOKY BAT This small, secretive bat is slate-gray to dark brownish gray. It generally roosts in well-lit areas inside rock crevices, small caves, tunnels, and buildings. The small colonies, found along the coastal regions from southwestern Ecuador to Chile, are easily scattered when the roost is disturbed by intruders.

107

DISK-WINGED BATS
Family Thyropteridae

There is only one genus and two species of these fascinating and delicate little mammals. Disk-winged bats are characterized by circular suction disks at the base of the thumb and on the ankle. These are specializations used for clinging to leaves while roosting. Their fur is long, soft, and chocolate to reddish brown above. Underneath they are whitish or the same color as above, depending on the species. Disk-winged bats have a head and body length of about 1½ inches. Their ears are broad at the base and have sharply pointed tips. The tail is longer than the legs, and all but the tiny tip is enclosed within the tail membrane. Both species feed on insects. These bats are found in lowland rainforest habitats from southeastern Mexico to Peru and Brazil.

TRICOLORED DISK-WINGED BAT These bats roost on the inside of banana or *Heliconia* leaves when the leaves are young and curled lengthwise, with an opening at the top. Moist suction disks are used to cling to the inside walls of these leaves while the bat roosts with its head facing up rather than down, as in most bats. Colonies of about six bats move together to new roosts every other day or so as the leaves mature and uncurl. The patchy distribution of these bats reflects the limited availability of their roost plants. Maternity colonies frequently roost in hollow logs. The bat's name refers to its coloring: dark to reddish brown on the top, lighter brown on the sides, and white or cream-colored underneath. Tricolored Disk-winged Bats, which range from southern Mexico to southeastern Brazil, eat beetles and flies.

**Tricolored Disk-winged Bats;
to 1½ in. (40 mm); leaving their coiled-leaf roost.**

OLD WORLD SUCKER-FOOTED BATS
Family Myzopodidae

This family is represented by a single species, called the Old World Sucker-footed Bat, and is found only on the island of Madagascar. This bat resembles disk-winged bats in having a high, domed crown and suction pads on their wrists and feet. The suction pads in the two families are rather different in detail and probably evolved independently. The ears of sucker-footed bats are large and shaped like a long leaf, with an unusual mushroom-shaped structure at the base on the inside. The tragus is attached along the edge of the ear. As in the smoky bats, the thumb and claw are poorly developed.

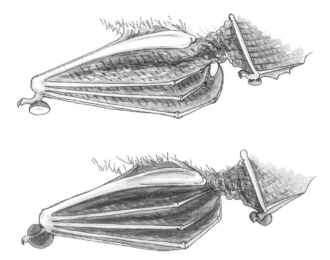

Comparison of feet and thumbs of Tricolored Disk-winged Bat *(above)* and Old World Sucker-footed Bat *(below)*.

Old World Sucker-footed Bat; to 2¼ in. (55 mm).

The biology of this bat is not well known. Like the disk-winged bats, sucker-footed bats use plants for roosts by clinging to the inside surfaces of rolled leaves and palm leaves. It is generally believed that they are restricted to rainforest habitats, where such plants are common. This rare bat, like most of the animal life in the Madagascar rainforest, is threatened and should be considered endangered.

PLAIN-NOSED BATS
Family Vespertilionidae

Plain-nosed bats are found throughout the world, except in the polar regions, and are the most frequently observed bats in many places. The family Vespertilionidae is more widespread and diverse than any other bat family. The rodent family Muridae, with about 1,000 species, is the only mammal family with more species. A plain nose, coupled with a prominent tragus (ear flap) and an interfemoral membrane that encloses the long tail, distinguishes vespertilionid bats from most others. Species in the families Natalidae and Furipteridae have funnel-shaped ears and poorly developed thumbs. Almost all vespertilionid species are insect eaters, but a few also eat fish or flower parts.

MOUSE-EARED BATS This genus, *Myotis,* contains almost 100 different species and is the most diverse of all bat genera. Commonly found in all but the coldest and most remote places, mouse-eared bats have the most extensive geographical distribution of any genus of land mammal except humans. Most are dark, are less than 3 inches long, and have a tragus that is pointed and erect.

LITTLE BROWN BAT This bat, belonging to the genus *Myotis,* is one of the most abundant species within its range, which extends from Alaska and Canada south to Mexico and the southwestern United States. It is also one of the best studied. As very efficient predators of nocturnal insects, Little Brown Bats benefit humans. A single bat can catch and consume more than 500 mosquitoes in an hour! When nursing their young, females eat up to their own weight in insects every night. Little Brown Bats often enter buildings during the summer without invitation. The warm environments in the attics are particularly favorable for the rapid growth of young. To prevent them from returning, you can plug the entrances after the bats migrate in the fall.

Bats may fly up to 200 miles to find a suitable hibernation site. Mating occurs in the fall, but sperm is stored in the female until the spring, when fertilization takes place. Hibernation may last six to eight months. Some individuals can live for more than 30 years in the wild. One southwestern subspecies is being considered for addition to the U.S. list of endangered species.

Little Brown Bat; to 2 in. (50 mm); drinking in flight.

GRAY BAT This handsome little bat used to be one of the most abundant mammals within its range, with huge columns of Gray Bats once literally darkening the evening skies. Unfortunately, Gray Bat populations have declined significantly over the past 20 years, and the Gray Bat is now listed as endangered by the U.S. Fish and Wildlife Service. Gray Bats are particularly vulnerable because they live in caves throughout the year, and 95 percent of them hibernate in only nine caves. These caves are deep, with large, cold rooms. Bats gather in large clusters in these rooms, where temperatures average 42° to 52° F. Females form maternity colonies during the summer, mostly in large, warm caves that contain streams or pools of standing water. These colonies are typically within a mile or so of a river or lake. Because of such specific habitat requirements, less than 5 percent of the available caves are suitable for Gray Bats. Fortunately, many of the occupied caves are now protected, and colonies have begun to recover. In fact, anyone who is interested can once again view the evening departure of Gray Bats from two maternal caves from May through September (see Batwatching, p. 146).

This huge gate restricts human entry into Hubbard's Cave in Tennessee. It protects one of the largest hibernating colonies of Gray Bats.

Merlin D. Tuttle, Bat Conservation International

Gray Bat (actual size); to 2 in. (50 mm); about to take off.

Gray Bats, which range from the cave regions of Missouri and Kentucky to northern Florida, are distinguished by their fur, which is uniformly gray from base to tip. The wing attachment in these bats is unique among North American *Myotis* bats in that it is on the ankle rather than at the base of the toes. Gray Bats typically forage for insects no more than 15 feet above the surface of rivers and lakes. Mayflies make up a large part of their diet, but they also consume the flying adults of many other aquatic insects.

Indiana Bat (actual size); to 1¾ in. (45 mm).

INDIANA BAT Although this species has been listed as endangered by the U.S. Fish and Wildlife Service since 1973, populations declined by approximately 55 percent between 1980 and 1988. Disturbances by cave visitors were responsible for many of the early losses. Indiana Bats hibernate in caves where temperatures average 38° to 43° F and the humidity is high. They form tightly packed clusters of about 300 bats per square foot. If disturbed while hibernating, bats may use up much of the fat they must have in order to survive hibernation. More than 85 percent of the remaining Indiana Bats—fewer than 400,000—hibernate in only seven U.S. caves and mines from Oklahoma to Vermont and northern Florida.

Despite efforts to protect them, many populations continue to decrease. The loss of habitat suitable for maternity colonies could be contributing to the population declines. The first Indiana Bat maternity roost was not discovered until 1974. This colony, and most of the others found thus far, were under the bark of trees located close to streams and rivers in or near floodplain forests. Bats establish foraging territories in these habitats, where they consume several kinds of insects, especially moths. They exhibit a high degree of loyalty to these localities, and each summer most bats return to the same areas.

Indiana Bats are usually dark gray, sometimes brownish. They differ from Little Brown Bats by having dull rather than glossy fur, a pinkish rather than brownish nose, and short rather than long hairs on the toes. Females form small maternity colonies where each mother bears a single pup in June or July.

There are 12 additional species of mouse-eared bats in the United States. Two of these are now being considered for addition to the U.S. list of endangered species. In Europe, ten species of Myotis are now either endangered or threatened. The reasons for the decline include roost disturbance, loss of roosting and foraging habitats, direct killing of bats, and indirect killing by pesticide poisoning. Chemicals used to treat timber against insect pests are implicated in a few of the European bat declines.

Gene Gardner

Northern Red Oak used by Indiana Bats as maternity roost.

SILVER-HAIRED BAT Silky black fur tipped with light silver makes this one of the most attractive bats in its range, which stretches from Alaska and southern Canada through most of the United States into northern Mexico. Silver-haired Bats differ from Hoary Bats (p. 128) in that they are smaller, they have only a sparse covering of fur on their tail membrane, and they have silkier fur.

Silver-haired Bats are most abundant during the summer in the coniferous forests of the Rocky Mountains and the northern parts of the United States and adjacent Canada. They forage near woodland ponds and streams for a variety of insects, including moths, bugs, beetles, flies, and caddis flies. They forage later in the evening than most other species. Their erratic flight is one of the slowest of North American bats. The loose bark of a tree often serves as a daytime shelter, but these bats may also roost in tree hollows or foliage.

Northern populations apparently migrate for the winter to the southern United States and northeastern Mexico. During their migration flight a few individuals may become disoriented and turn up on ships or islands distant from the mainland. Silver-haired Bats occasionally migrate with Red Bats and small birds. During migration all three groups frequently collide with radio towers and high buildings. Silver-haired Bats apparently have a well-developed homing sense and make it safely back home. Some populations may hibernate locally during the winter, with single individuals or small groups moving into buildings, trees, cliffs, caves, and mines. Females usually bear one or two young in June or July.

Silver-haired Bat;
to 2¼ in. (60 mm);
seen here emerging
from roost under loose bark.

PIPISTRELLES This large genus of about 43 species is widespread and can be found throughout much of the world except South America. Pipistrelles differ from mouse-eared bats by having ears that are usually shorter and broader and a tragus that has a rounded, rather than a pointed, tip. These bats are some of the first to appear in the evening and can be identified by their erratic flight. Very large colonies of the Common Pipistrelle have disappeared altogether in Europe, where agricultural pesticides and chemicals used to treat timber have also killed many other bats.

Eastern Pipistrelles can be found in the forests of eastern North America and Central America. Western Pipistrelles are found in desert regions of the western United States and Mexico. Western Pipistrelles are predominantly gray, whereas Eastern Pipistrelles are usually yellowish brown, with fur that has three distinct bands of color. Western Pipistrelles are the smallest bats in North America, with a head and body length of about 1½ inches. The northern populations of both of these species hibernate in caves, mines, and rock crevices. Both species produce twins from late May to early July.

Eastern Pipistrelle; to 1½ in. (40 mm); awakened briefly during hibernation. Water droplets condense on the bat as it sleeps, and its body temperature drops to that of the cave.

Big Brown Bat; to 2¾ in. (70 mm).

BIG BROWN BAT People encounter this species, also called the House Bat, more often than any other North American species because it can live in buildings year-round. Big Brown Bats also roost in caves, hollow trees, tunnels, rock crevices, and even hollow cactus trunks. In the U.S., this species is distinguished by its relatively large size, wide nose, and broad, rounded tragus.

Big Brown Bats have an unusual tolerance for cold and can survive body temperatures below freezing. This accounts for their ability to spend the winter in places as far north as Canada. Individuals may then travel 150 miles to hibernate in a cave or a mine. Hibernating bats may become active and change roosts during unseasonably warm or cold winter weather. They have even been seen flying in snowstorms! Beetles are the primary component of their diet, which also includes flying ants and flies. Mating occurs in the fall, and females produce one or two young in late May or early June. Big Brown Bats can be found from North America through Central America to northern South America.

Greater Bamboo Bat; to 1¾ in. (45 mm).

BAMBOO BATS Their small size and extremely flattened head and body, as well as disk-shaped cushions on the thumb and foot, make these bats easy to identify. Bamboo bats are reddish or dark brown above and paler below. Their size and shape are adaptations that allow them entry into the hollow joints of bamboo through holes created by beetles. The cushions enable the bats to cling to the interior walls of the bamboo. Harems or family units of a dozen or more may use a single section of bamboo. Greater Bamboo Bats are found from India and China to southern Indonesia. They have twins in April or May.

EVENING BAT This species, found from North America through Central America and the Caribbean to northern South America, looks like the Big Brown Bat but is smaller and has two, rather than four, upper incisors. The teeth, plus a tragus that is short, curved, and round, distinguish Evening Bats from other small brown bats in the United States. Evening Bats are abundant in the southern parts of their range, where populations remain active year-round. Northern populations apparently migrate to warmer, southern regions in the winter.

Evening Bats usually roost in buildings but are also found behind loose bark, in tree hollows, inside clumps of Spanish moss, and under palm fronds. They almost never enter caves. Harems may be present in some populations. Maternity colonies numbering into the hundreds are formed in spring; females give birth to twins and occasionally triplets. Each mother nurses her young until the pups are about a month old. Pups may occasionally nurse from a female other than their mother after that. They begin flying within three to four weeks of birth.

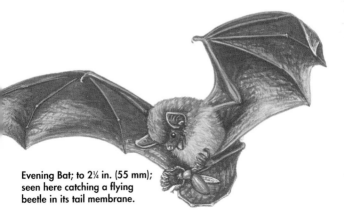

Evening Bat; to 2¼ in. (55 mm); seen here catching a flying beetle in its tail membrane.

Variegated Butterfly Bat; to
2½ in. (63 mm); range is
West Central Africa.

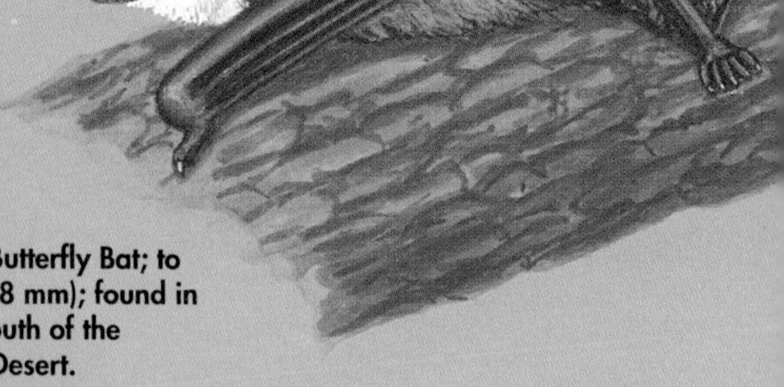

Superb Butterfly Bat; to
2¼ in. (58 mm); found in
Africa south of the
Sahara Desert.

**Asian Yellow House Bat; to 2 in. (50 mm);
range is Pakistan and Southeast Asia to Java.**

BUTTERFLY BATS These attractive bats have a rather slow, butterflylike quality to their flight and often white spots and stripes on their back on a background of cream, brown, gray, or black. A fleshy lobe at the base of the mouth narrows as it nears the bottom of the ears. All ten species of butterfly bats are found in tropical rainforests and bushy savanna habitats in Africa southward from the Sahara to Namibia. Palm fronds, banana leaves, bushes, tree holes, and primitive huts serve as roosts. The bats' unusual coloring may provide camouflage in these exposed roosts. Butterfly bats often begin to hunt for insects in daylight. One or two young are born in early spring.

YELLOW HOUSE BATS As the name implies, these bats are usually yellow (though some may be brown), and they frequently live in houses. Yellow house bats are heavy-bodied, strongly built bats with a very long tragus that is curved forward. Nine species can be found in Madagascar and in Africa south of the Sahara and from Pakistan to the Philippines and Indonesia. They prefer to roost in the warm attics of houses with metal roofs, but they will also use hollow trees, woodpecker holes, and palm fronds on occasion. Yellow house bats feed on beetles, termites, moths, and other insects. Females may form maternity colonies, and all bear one or two young.

125

Eastern Red Bat (male); to 2¼ in. (55 mm); seen here sleeping in its usual habitat, the foliage of a deciduous tree.

EASTERN RED BAT This striking bat, found from central and eastern Canada and the United States into Central America, has long, frosted fur that is bright reddish orange to reddish brown. Males are usually more brightly colored than females. As with other tree-dwelling bats in the genus *Lasiurus,* the interfemoral membrane is heavily furred and extends straight backward during flight. A newly recognized and closely related species, called the Western Red Bat, occurs from southwestern Canada to southern South America. This bat is darker and lacks the frosting of the eastern species. Seminole Bats are about the same size, but they are a deep mahogany-brown. All other so-called tree bats in North America are larger.

Eastern Red Bats are found in forests and open areas with small patches of trees. When hanging by one foot from a leaf or stem, an Eastern Red Bat looks very much like a dead leaf. Trees usually serve as day roosts, but bushes along fence rows and even sunflower leaves are occasionally used. The area below the roost must be open enough for the bats to drop down and begin flight.

Eastern Red Bats emerge early in the evening and frequently can be observed foraging in elliptical patterns near forest edges, along tree-lined roads, and around streetlamps. They consume a wide variety of insects, particularly moths and leafhoppers.

In northern temperate areas, Eastern Red Bats migrate south for the winter, while those in warmer regions are permanent residents. During very cold weather, Eastern Red Bats do not arouse from torpor unless the temperature outside warms up enough for insects to become active. They use their furry tail membrane like a blanket to cover themselves, including their face, thereby minimizing heat loss. Eastern Red Bats mate in the fall, and young are born from May to early July. Mothers frequently give birth to triplets and even quadruplets—an unusually large number of young for bats.

Eastern Red Bat mother rescuing one of her young that has fallen to the ground.

Hoary Bat; to 3¼ in. (85 mm); in
defensive posture.

HOARY BAT This large, furry bat, with its short, round ears
outlined in black, looks like a small teddy bear. The frosty,
or hoary, appearance is caused by the white tips on the
yellow and dark brown hairs. The Hoary Bat has a yellow
throat and a heavily furred interfemoral membrane. It is
much larger than the Silver-haired Bat (p. 118), which also
has frosty fur. Other tree-dwelling bats in the genus
Lasiurus have yellowish to reddish fur.

The Hoary Bat is one of the most widespread of all
mammal species. It ranges from Canada and Iceland south
to Hawaii and Argentina. As is typical of tree-dwelling
species in North America, Hoary Bats roost alone in trees
and are most often seen in or near forests. They sometimes
visit caves during the late summer and early autumn but
almost never enter buildings. Long, thick fur insulates these

bats from extreme cold and makes it easier for them to fly in freezing conditions. Northern populations migrate southward, sometimes in waves, during the fall and return in the spring. Southern and tropical populations apparently do not migrate. Hoary Bats are among the fastest-flying bats, with a swift, direct flight that makes them easy to identify on the wing. They eat a variety of insects, but moths are preferred. Mother bats give birth to two babies in late spring or early summer. Pups nurse during the day and remain at the tree roost while mothers hunt for food. Adult male bats generally stay away while the pups are nursing.

The Hawaiian populations of this bat, known in Hawaii as O'pea'pea, represent a distinct subspecies. This is the only bat that is native to Hawaii, and it is on the U.S. list of endangered species.

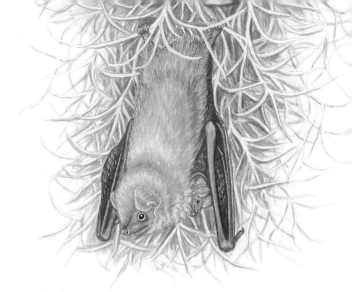

Northern Yellow Bat; to 3 in. (75 mm); mother with young.

OTHER NEW WORLD TREE BATS Seminole Bats are about the same size as Eastern Red Bats, but they are a deep mahogany-brown. The range of the Seminole Bat is similar to the distribution of Spanish moss, a favorite roosting place for this bat. This range includes the Gulf States from eastern Texas to North Carolina and southern Tennessee. Northern and Southern Yellow Bats are similar to each other in color. The latter species has a head and body length that is almost an inch shorter than that of the former. Northern Yellow Bats routinely roost in Spanish moss, from South Carolina to Honduras to Cuba, and bear from two to four young once a year. Southern Yellow Bats can be found from the southern United States southward to Uruguay and Argentina. They roost in leaves and palm fronds and usually have two young.

EUROPEAN BARBASTELLE The notched ears of this bat are unique. The black fur, which extends onto the tail membrane and wings, is long, soft, and tipped with white, producing a silvery appearance. European Barbastelles are rare throughout much of their range, which extends from Morocco and Europe to the southwestern part of what was formerly the Soviet Union. They begin their search for insects before sunset, usually over water or in woodlands close to water. During warm weather European Barbastelles roost on tree trunks, inside tree hollows, under bark and leaves, and in buildings. They migrate south during the winter to hibernate in caves. Entire colonies have been lost due to the chemicals used to treat lumber.

European Barbastelle; to 2 in. (50 mm).

Townsend's Big-eared Bat; to 2 in. (50 mm); seen here flying out of an abandoned mine.

TOWNSEND'S BIG-EARED BAT Populations are extremely low for at least two endangered subspecies of this bat—the Ozark and Virginia Big-eared Bats. The Ozark subspecies is restricted to eastern Oklahoma and adjacent Arkansas, whereas Virginia Big-eared Bats are found in parts of Kentucky, Virginia, West Virginia, and North Carolina. Another subspecies, which is found from northern California to southwestern Canada, may also be endangered. The primary reason for these declines is thought to be the human disturbance of roosts.

These pale brown to nearly black bats have unusually

long ears and possess two large glands above the nostrils. Some populations roost in buildings during the summer, but eastern populations are restricted to caves or mines in both summer and winter. Townsend's Big-eared Bats often roost singly but may form tight clusters of a few to a hundred or more bats during hibernation. Mating takes place in the fall or early winter, and a single young is born during May or June. Townsend's Big-eared Bats begin foraging well after dark and are very agile in their pursuit of moths.

OTHER BIG-EARED BATS FOUND IN THE UNITED STATES
Rafinesque's Big-eared Bat, closely related to Townsend's, has whitish rather than tan underparts, gray fur, and long hairs on the toes. This poorly known bat is found from Illinois southward to Texas and Florida. It is rare and a candidate for the endangered species list. Allen's Big-eared Bat, found in the southwestern United States and north central Mexico, has flaps of skin projecting over the nose. This species, which is also poorly studied, typically roosts alone in caves and mines. It emits loud calls audible to people.

Allen's Big-eared Bat (actual size); to 2¼ in. (60 mm); shown here hibernating. The ears are coiled to minimize heat loss.

SPOTTED BAT This is certainly one of the most striking species of bat in North America. It has three large white spots on a coat of black fur, pinkish ears that are almost as long as its body, and pink wings.

Spotted Bats also have distinctive echolocation calls. Scientists have described these calls as high-pitched, metallic "pings."

Spotted Bats are among the least known mammals in North America, and they may be rare throughout their range, which includes southwestern Canada and the western portion of the United States to Central America. These bats may eventually be added to the U.S. endangered species list.

Ponderosa pine forests appear to be the preferred habitats of Spotted Bats. There they can occasionally be heard foraging above the canopy.

Rocky deserts may be inhabited by these bats more often than pine forests during the late summer and early fall. Daytime roosts have been found in the cracks and crevices of high cliffs and broken canyon walls.

Spotted Bats are very fast flyers. Noctuids and other small moths, which are the Spotted Bats' preferred foods, are captured in the air as the bats forage alone. Occasionally they may pick insects off the ground or foliage. Individual Spotted Bats may establish their own foraging territories.

There are very few records on the winter activities and habits of Spotted Bats. It is not known whether they hibernate or migrate to warmer climates. A single pup is born in May or June.

Spotted Bat;
to 2½ in. (65 mm).

COMMON LONG-FINGERED BAT This species is also called the Common Bent-winged Bat. The names refer to the third finger, which is exceptionally long and is folded back upon the wing when the bat is roosting. The Common Long-fingered Bat is reddish brown to black and has a short nose and a high crown. In the fall, bats congregate at caves to mate, and young are born in late spring or early summer. Implantation of the fertilized egg is delayed for up to five months, a reproductive strategy unique among hibernating bats. Populations in cooler regions migrate and hibernate. Large maternity colonies may number into the tens of thousands. Common Long-fingered Bats also roost in rock crevices, trees, and buildings. Beetles and other insects comprise their diet. These bats are endangered in Europe. There are ten other species in this genus found in Africa and Madagascar as well as in parts of southern Europe, Asia, and Australia.

Common Long-fingered Bat (actual size); to 2½ in. (60 mm).

Orange Painted Bat; to 2 in. (50 mm);
ranges from India to southern China
and south to Java and Bali.

PAINTED BATS Striking and colorful patterns characterize several of the 16 species in this genus. Colors include orange or scarlet fur and black wings with orange stripes. Some species are reddish brown to gray, whereas other species have gray blotches, making them look somewhat like clumps of moss. Painted bats also possess long, woolly, rather curly hair and are sometimes called woolly bats. These small bats are from 1½ to 2 inches in head and body length. They roost singly or in small groups in trees, buildings, and bird nests, where they may be mistaken for dead leaves or a nest of wasps. One species has even been found roosting in a spider web! Painted bats are found from Africa and India through Asia, the Philippines, Indonesia, and Australia. They have a fluttery flight pattern and forage late in the evening for insects. In central Africa, young are born in October.

137

PALLID BAT This bat, found in southwestern Canada through the western United States into Baja California and central Mexico, is well adapted for desert environments. It has long and broad ears, large eyes, and a broad muzzle that lacks the distinctive facial bumps or flaps of other big-eared bats. Its color, yellow-brown on top and cream below, is difficult to see against a rocky background and thus may provide camouflage from predators. Pallid Bats are efficient at retaining body water. Adults can live for up to a month solely on water derived from insects. Their large ears are good at detecting sounds that insects and other arthropods make as they move along the ground. Favorite foods include crickets and grasshoppers, beetles, moths, and scorpions. They eat their prey on the ground or take it to a night roost.

During the summer small colonies often roost in wide horizontal rock crevices that remain warm both day and night. These roosts provide a suitable environment for the developing young and are safe for pups. In colder periods vertical crevices are used because bats can move up and down to cool or warm themselves. Shallow caves, cliffs, and buildings are often selected as night roosts by clusters of bats.

Mating occurs in late fall, and most individuals hibernate during winter. Maternity colonies are formed and babies are born in May or June. Pallid Bats can have from one to three young, but they usually bear twins. Female roost mates have been observed helping each other through labor and helping to locate lost young. They will also baby-sit for each other's pups. Young that are able to fly are taught roost locations and may remain in the group for more than a year. Although at least six distinct communication calls are used by Pallid Bats, vision and smell are also important.

Pallid Bat; to 2½ in. (65 mm);
seen here snatching scorpion from desert floor.

FREE-TAILED BATS
Family Molossidae

The long, narrow, leathery wings of the members of this family are designed for rapid and prolonged flight. The name "free-tailed" refers to the characteristic tail that extends far beyond the trailing edge of the interfemoral membrane. However, the membrane can be extended to the tip of the tail when it is being used to trap insects in flight. Free-tailed bats have a rather wide head and muzzle. Their short legs are strong, and their feet are broad. Bristles of hair are scattered on their head and feet. A variety of roosts are used, including caves, rock crevices, tunnels, buildings, hollow trees, loose bark, foliage, holes in the ground—even rotting logs. Cave colonies may number into the millions, with the bats in these colonies producing tremendous quantities of droppings. Some species migrate locally or seasonally, but none hibernates. Free-tailed bats usually have a single pup every year.

Long-crested Free-tailed Bat; to 2¼ in. (55 mm); found in Africa.

Merlin D. Tuttle, Bat Conservation International

Naked Bat; to 5 in. (130 mm);
seen here crawling in cave.

NAKED BAT Hunting by humans has destroyed all but a few populations of this remarkable bat, found on the Malay Peninsula and the Philippines south to Java. It is large and naked except for scattered short hairs and a patch of bristles near a highly developed throat sac. The black or dark gray skin of Naked Bats is thick and pliable. A thumblike toe on each foot is used to push the wings into skin pouches on the sides of the body so that the forearms can be used for moving on the ground. The wide muzzle is accentuated by large, smooth lips. Naked Bats roost in large caves or hollow trees. They forage for insects over streams and clearings and high over the forest.

MEXICAN FREE-TAILED BAT Probably more people have observed this bat than any other. Mexican Free-tailed Bats are the famous bats of Carlsbad Caverns in New Mexico and of the Congress Avenue Bridge in Austin, Texas, where millions of people have witnessed their impressive evening flights. With a body mass of about half an ounce, Mexican Free-tailed Bats are the smallest molossid species in the United States. They are also the only species in which the ears do not join and in which the hair is uniformly colored. The short, dense fur is grayish brown above and lighter-colored below.

There is considerable variation in the behavior and ecology of Mexican Free-tailed Bats across the United States. Populations between Sierra Nevada and eastern Texas form huge maternity colonies with several million bats in large caves. Colonies may also form under bridges and in buildings. The females and young go to Mexico during the winter. Most males do not migrate; instead, they remain in Mexico throughout the year. Populations to the east and west form smaller colonies and do not migrate. Eastern populations typically roost in buildings, under bridges, and in hollow trees. Populations in Latin America have not been studied in depth.

Mexican Free-tailed Bats feed principally on moths but may also eat beetles and flying ants. A colony of 20 million bats may consume almost half a million pounds of insects each night! Physiologically, these bats are capable of dealing with the high levels of ammonia and carbon dioxide frequently encountered in large roosts. A single pup is born in June or July. Mothers are able to locate and nurse their own young, even in large colonies where pups cover the walls in densities of up to 500 per square foot!

Many populations, including the one at Carlsbad

Caverns, have declined during the past several decades. Disturbances at the bat roosts, vandalism, and agricultural pesticides have been the major causes. In just six years, vandalism apparently destroyed almost all 30 million Mexican Free-tailed Bats that lived in Eagle Creek Cave, Arizona, once the world's largest colony of bats.

Big and Pocketed Free-tailed Bats are seen much less frequently in the western United States and Mexico. Their fur is bicolored, and their ears are joined at the base.

Merlin D. Tuttle, Bat Conservation International

**Mexican Free-tailed Bat; to 2¼ in. (60 mm);
mother surrounded by her babies.**

MASTIFF BATS All three species of mastiff bats in the United States are being considered for addition to this country's list of endangered species. These bats differ from the other molossid genera in the United States by being larger and having smooth lips. Their ears are very large and project forward almost to the point of covering the eyes. These bats roost in rock crevices, tunnels, trees, and under the heat-trapping metal roofs of buildings. Roosts are usually 15 feet or higher above the ground. This allows room for these large, narrow-winged bats to become airborne. As they fly and feed on insects, mastiff bats emit high-pitched "peeps" audible to humans.

With a wingspan of 21 inches, the Western Mastiff Bat, found in the southwestern United States to Argentina, is the largest bat in North America. Underwood's Mastiff Bat is almost as large but can be distinguished by its smaller ears, which barely reach to the tip of the nose when laid forward. The latter species is found from Arizona to northern Central America. Wagner's Mastiff Bat ranges from the southern tip of Florida to Cuba and southern Mexico southward through Central and South America to Uruguay. Mastiff bats are quite rare in the United States and apparently do not migrate.

Western Mastiff Bats;
to 4¼ in. (110 mm).

BATWATCHING

Imagine huge clouds of bats leaving a roost, seen against the backdrop of a summer sunset. This is one of the most dramatic wildlife spectacles offered by nature. The list below provides a few of the best places to watch bats in the United States and its territories. Foraging bats can be seen over almost any body of water during warm summer nights. You may be able to locate them by eavesdropping on their echolocation calls with the aid of an ultrasonic bat detector. This device converts ultrasonic frequencies to sounds that are audible to humans. Captive bats offer an opportunity to get to know bats "up close and personal," and many zoos now have bat exhibits.

Selected Batwatching Sites

Alabama: Blowing Wind Cave (Wheeler National Wildlife Refuge) and Hambrick Cave (Tennessee Valley Authority), endangered Gray Bat.

American Samoa: Proposed Samoan National Park, Samoan and Tongan Flying Foxes.

California: Point Reyes National Seashore, 14 species, including endangered Townsend's Big-eared Bat.

Missouri: Rock Bridge Memorial State Park, endangered Gray and Indiana Bats.

New Mexico: Bandelier National Monument and Carlsbad Caverns National Park, 12 species, especially Mexican Free-tailed Bat.

Tennessee: Nickajack Cave (Tennessee Valley Authority), endangered Gray Bat.

Texas: Congress Avenue Bridge in Austin, Eckert James River Bat Cave (Texas Nature Conservancy), and Old Tunnel Wildlife Management Area, Mexican Free-tailed Bat.

You can contact Bat Conservation International, P.O. Box 162603, Austin, TX 78716 (512-327-9721), for more information about bat detectors, any of these sites, and other batwatching opportunities.

One of the most famous batwatching sites in the world, Congress Avenue Bridge in Austin, Texas.

Merlin D. Tuttle, Bat Conservation International

YOU CAN INVITE BATS to your backyard and become active in bat conservation by using an artificial roost to attract bats. Bat houses are enjoyed throughout the United States by a variety of bat species that also use natural crevices and tree hollows. Even if the house is not used by bats, its presence causes visitors to ask about the house, which provides the owner an excellent opportunity to educate friends about bats! Here's how to build your own bat house:

MATERIALS NEEDED:

One 9′ long piece of 1″ x 8″ lumber, cut into 6 pieces, each 15¾″ long for front and back and entrance restriction.

One 5′ long piece of 2″ x 2″ lumber, cut into 2 pieces, each 22″ long for sides, and one piece 12¾″ long for ceiling.

One 16½″ long piece of 1″ x 4″ lumber for roof.

One piece of 15½″ x 23″ fiberglass screening (do not use metal screening).

PROCEDURE:

1. Nail back to sides and ceiling.
2. Staple fiberglass screening to inside surfaces of back and sides.
3. Nail entrance restriction to front, then nail front and roof to sides and ceiling.
4. Make sure all surfaces are free of sharp points.
5. Hang house with hooks or nails about 15′ or higher on a pole, house, or tree. Make sure that the entrance area is clear of obstructions.
6. The probability of occupancy by bats is increased if the house is placed within a quarter of a mile from a stream, river, or lake and if it receives appropriate sunlight.
7. Take notes of your occupancy patterns and share your results with neighbors and with Bat Conservation International's Bat House Research Project.

BAT HOUSE PLANS

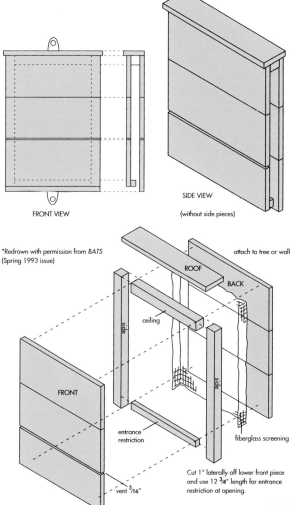

FRONT VIEW

SIDE VIEW

(without side pieces)

*Redrawn with permission from *BATS* (Spring 1993 issue)

attach to tree or wall

ROOF

BACK

ceiling

side

side

FRONT

entrance restriction

fiberglass screening

vent $^{3}/_{16}$"

Cut 1" laterally off lower front piece and use 12 $^{3}/_{4}$" length for entrance restriction at opening.

KEEPING BATS OUT of your home is not difficult. A safe and humane solution to keeping bats out that have moved into your attic or behind your walls is to watch to see where bats leave the building at dusk, then identify the specific holes or cracks during the day by looking for brownish stains at the exit sites. These sites will frequently be found where a wall meets a roof or a chimney, behind a loose board, between the slats of an attic vent, in holes around windows or doors, or in holes made by other animals. Wait until the young can fly or the bats are due to migrate, then hang bird netting over the exit points, allowing it to hang at least 2 feet below the exit site. The bats can crawl out from under the loose netting but cannot find their way back in. Netting can be purchased from garden centers and some hardware stores. Three days after the netting has been in place, close over the holes with screening, boards, caulking, or similar material.

Ultrasonic devices and chemical repellents are either ineffective or effective for only a short period of time. The use of poisons, such as the tracking powder called Rozol, and other chemicals should be avoided altogether because of the severe health problems, including death, they can cause in humans. In addition, these poisons are costly and have to be reapplied frequently.

ORGANIZATIONS TO HELP PROTECT BATS do exist. And although some bats are endangered, their situation is far from hopeless. In fact, more and more people, in private organizations and government groups, are becoming concerned with the conservation of bats. This increased awareness, as well as action, is largely due to the efforts of Bat Conservation International (BCI), the Chiroptera Specialist Group of the International Union for the Conservation of Nature and Natural Resources, the Flora and Fauna Preservation Society, the National Speleological

Society, the World Wildlife Fund, and others. All of these organizations emphasize education, which is the key to bat conservation in the future.

BCI has become a world leader in bat conservation efforts, with programs in education, management, habitat protection, and conservation legislation. The Nature Conservancy has taken action to protect numerous bat caves in the United States, including several that harbor some of the largest hibernating populations of two endangered species. Many state wildlife agencies have taken important steps to study and manage local populations of bats. The United States government has increased its bat conservation efforts by adding species to the list of those endangered, by passing legislation to protect caves on federal lands, and by promoting international efforts to limit the trade in Pacific species of flying foxes. Island nations in the Pacific have also implemented laws regulating the hunting and commercial trade of flying foxes. Reserves have been established in England, Poland, and other European countries to protect important bat roosts. Most of these countries recently ratified two international laws protecting European bats. Finally, the Lubee Foundation in Florida, the Jersey Wildlife Preservation Trust in England, and several U.S. zoos have started captive breeding programs for endangered species of bats.

FOR MORE INFORMATION

BOOKS

Barbour, R. W., and W. H. Davis. *Bats of America*. University Press of Kentucky, Lexington, KN, 1969.

Fenton, M. B. *Bats*. Facts on File, Inc., New York, NY, 1992.

———— *Communication in the Chiroptera*. Indiana University Press, Bloomington, IN, 1985.

Fenton, M. B., P. Racey, and J. M. V. Rayner. *Recent Advances in the Study of Bats*. Cambridge University Press, Cambridge, England, 1987.

Fleming, T. H. *The Short-tailed Fruit Bat*. University of Chicago Press, Chicago, IL, 1988.

Geluso, K. N., J. S. Altenbach, and R. C. Kerbo. *Bats of Carlsbad Caverns National Park*. Carlsbad Caverns Natural History Association, Carlsbad, NM, 1987.

Griffin, D. R. *Listening in the Dark*. Yale University Press, New Haven, CT, 1958.

Hill, J. E., and J. D. Smith. *Bats: A Natural History*. University of Texas Press, Austin, TX, 1984.

Kunz, T. H. *Ecological and Behavioral Methods for the Study of Bats*. Smithsonian Institution Press, Washington, D.C., 1988.

———— *Ecology of Bats*. Plenum Press, New York and London, 1982.

Mickleburgh, S., A. M. Hutson, and P. A. Racey. *Old World Fruit Bats: An Action Plan for Their Conservation*. IUCN/SSC Chiroptera Specialist Group, Gland, Switzerland, 1992.

Morton, P. A. *Educator's Activity Book About Bats*. Bat Conservation International, Austin, TX, 1991.

Novick, A., and N. Leen. *The World of Bats*. Holt, Rinehart, and Winston, New York, NY, 1970.

Stebbings, R. E. *Conservation of European Bats*. Christopher Helm, London, 1988.

Tuttle, M. D. *America's Neighborhood Bats*. University of Texas Press, Austin, TX, 1988.

Tuttle, M. D. , and D. L. Hensley. *The Bat House Builders Handbook*. Bat Conservation International, Austin, TX, 1993.

Wilson, D. E., and G. L. Graham. *Pacific Island Flying Foxes: Proceedings of an International Conservation Conference*. U.S. Fish and Wildlife Service, Biological Report 90(23), 1992.

Wimsatt, W. A. *Biology of Bats*. Vols. 1, 2, 3. Academic Press, New York, NY, 1970.

MAGAZINES AND NEWSLETTERS

BATS. Quarterly of Bat Conservation International since 1982.

Bat News. Quarterly of the Fauna and Flora Preservation Society.

SCIENTIFIC NAMES

The common names of the bat species discussed in this book are listed alphabetically; their scientific names follow in *italics*. **Heavy type** indicates text pages. Most common names were taken from existing references, but a few were created because there isn't any accepted common name.

The scientific name consists of two parts: the genus to which the species belongs and the name that identifies the particular species within the genus. An asterisk (*) indicates that all or most of the species in this genus are grouped here under a single common name.

155

INDEX

Italicized page numbers refer to illustrations.

159